OTFS: Orthogonal Time Frequency Space Modulation
A Waveform for 6G

RIVER PUBLISHERS SERIES IN COMMUNICATIONS

Series Editors:

ABBAS JAMALIPOUR
The University of Sydney
Australia

MARINA RUGGIERI
University of Rome Tor Vergata
Italy

JUNSHAN ZHANG
Arizona State University
USA

The "River Publishers Series in Communications" is a series of comprehensive academic and professional books which focus on communication and network systems. Topics range from the theory and use of systems involving all terminals, computers, and information processors to wired and wireless networks and network layouts, protocols, architectures, and implementations. Also covered are developments stemming from new market demands in systems, products, and technologies such as personal communications services, multimedia systems, enterprise networks, and optical communications.

The series includes research monographs, edited volumes, handbooks and textbooks, providing professionals, researchers, educators, and advanced students in the field with an invaluable insight into the latest research and developments.

For a list of other books in this series, visit www.riverpublishers.com

OTFS: Orthogonal Time Frequency Space Modulation

A Waveform for 6G

Suvra Sekhar Das

Indian Institute of Technology Kharagpur, India

Ramjee Prasad

Aarhus University, Denmark

River Publishers

Published, sold and distributed by:
River Publishers
Alsbjergvej 10
9260 Gistrup
Denmark

www.riverpublishers.com

ISBN: 978-87-7022-656-1 (Hardback)
   978-87-7022-655-4 (Ebook)

©2021 River Publishers

The authors dedicate this book to the Nation.

Contents

Preface

कर्मण्येवाधिकारस्ते मा फलेषु कदाचन ।
मा कर्मफलहेतुर्भूर्मा ते सङ्गोऽस्त्वकर्मणि ।।

karmaṇy-evādhikāras te mā phaleṣhu kadāchana
mā karma-phala-hetur bhūr mā te saṅgo 'stvakarmaṇi

(Bhagavad Gita, Chapter 2, Verse: 47)

Translation

You have the right to work only but never to its fruits.
Let not the fruits of action be your motive, nor let your attachment be to inaction.

We are at a time where 5G is getting launched all across the globe. It's appropriate time to look forward to the next generation viz. 6G.

Most of the generations namely 1G,2G, 3G and 4G were distinct from each other in one common area of technology development namely the Air interface / Waveform / Transmission technology. While 1G had analog waveform, 2G had digital time division multiplexing, 3G had code division multiple acces and 4G used Orthogonal frequency division multiplexing (OFDM). The fith generation (5G) was an extension of OFDM with flexible parameters.

During the run up to 5G, several waveforms were investigated however due to several techno-commercial reasons the simplified flexible numerology based OFDM was elected as the waveform of choice in order to support higher mobility requirements while allowing for backward compatibility with 4G.

While OFDM allows for a low complexity transceiver, it is well known that OFDM is designed for linear time invariant channels, whereas high mobility and higher carrier frequency of operations lead to higher intercarrier interference (ICI) owing to Doppler and phase noise.

The different waveforms developed during the early investigation stage of 5G had several desirable properties in line with the initial requirement of 5G. In a sequence of developments, orthogonal time frequency space modulation (OTFS) also appeared soon after the flexible numerology based OFDM was elected as the waveform for 5G.

OTFS is inherently designed to operate in highly time variant channels. In this book we describe the foundation of OTFS as well as present realizable transceiver architectures and algorithms along with its extension mutiple access schemes.

Kharagpur, India.   *Suvra Sekhar Das*
May 2021   *Ramjee Prasad*

Acknowledgements

The authors are immensely grateful to Dr. Shashank Tiwari, Dr. Aritra Chatterjee and Mr. Vivek Rangamari, who have put in a great amount of effort in laying the foundation of this manuscript without which the book would not have seen the light of the day. It was due to their tireless effort that the book could get started and finally get completed. The authors are highly grateful to all members of IIT Kharagpur, with a special thanks to Prof. Saswat Chakrabarti, G.S. Sanyal School of Telecommunications and Prof. V. K. Tiwari, director IIT Kharagpur, for providing the necessary support and facility to conduct research work, only because of which the contents in the books could have been produced.

The first author would like to express his gratitude to his parents, Mr. Chandra Sekhar Das, Mrs. Anjali Das, because of whose sacrifices and effort he is able to contribute this little gift to the future researchers of wireless communications. The first author is indebted to his sister, Anusua whose affection has always been with him. He is also greatly thankful to his wife Madhulipa and two children Debosmita and Som Sekhar for bearing through the tough period during the preparation of this work. It is extremely important to highlight the contributions of the various researchers who have made the book what it is.

The second author acknowledges the love and affection his grandchildren Sneha, Ruchika, Akash, Arya and Ayush have showered on him.

List of Figures

List of Tables

1

Introduction

This chapter presents a background on the development of mobile communication systems over different generations.

1.1 Background

A mobile communication system consists of two separate parts, one is the core network which carries traffic from one edge network node at one end to another edge network node at the other end of the telecommunication system. The other part is the radio access network (RAN), which enables users to connect to the network using radio modems.

The core network already existed because of plain old telephone service (POTS) systems, whereas the radio access part is necessary to provide the last mile wireless mobile connectivity. Providing wireless connectivity has its own challenges such as power dissipation loss in comparison to wired connectivity. However, major challenges arise while providing connectivity to highly mobile user terminals as well as in multipath propagation terrains where excess delays become large. The evolution of different generations of wireless communication systems is primarily due to advancement in the air interface or radio access technology in the RAN which has been developed in order to meet the ever increasing demand to support high data rate, low latency, high reliability in adverse propogation condition. However, the recent fifth generation (5G) networks involve significant changes in core network architecture as well. In order to meet these requirements, the methods which received attention include forward error correction codes, multi-antenna technology, and channel state information (CSI) feedback–based link adaptation and resource allocation. Whereas, one of the very important attributes which distinguishes these generations is the waveform design. We briefly trace the evolution of waveforms below, while a summary of the waveforms is provided in Figure 1.1.

1.2 1G - 2G

First-generation (1G) communication systems were analog in nature, which used narrow-band frequency modulation having 30 KHz bandwidth [3]. Although 1-G

1

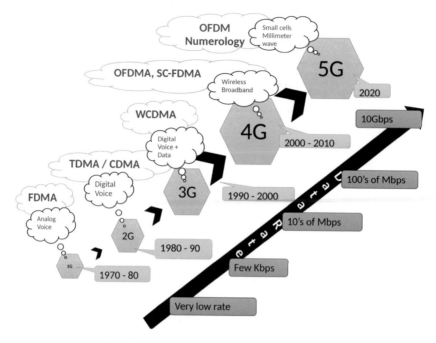

Figure 1.1   Evolution of waveform from 1G to 5G.

communication systems were designed for coverage and voice service, different 1-G systems were incompatible with one another, which led to standardization efforts such as Groupe Speciale Mobile (GSM) in Europe. The outcome of such efforts was second-generation (2G) digital communication systems, which used constant modulus Gaussian minimum shift keying (GMSK) modulation [4] with 200 KHz system bandwidth. High range of communication was enabled in 2G by GMSK as it has low peak to average power ratio (PAPR) and very low out of band (OoB) radiation, which reduce adjacent channel interference. The 2G signal suffers from inter-symbol interference, which is equalized using maximum likelihood sequence estimation (MLSE) receivers [5].

1.3 2G - 3G

Although communication upto a data rate of 473 Kbps was enabled in 2G by the introduction of general packet radio services (GPRS) and enhanced data for global evolution (EDGE), the third-generation (3G) communication systems were envisioned to provide high data rates up to 14 Mbps. In order to support such high data rate, a system bandwidth of 5 MHz was needed to be used, which is significantly larger than the 200 KHz bandwidth used in earlier generation systems. Wide-band code division multiple access (WCDMA) modulation was chosen for 3G due to its immunity to narrow-band interference and inherent security capabilities owing to its noise like power spectral density. The effect of multi-path fading was compensated through receiver architecture following rake structure [6].

1.4 3G - 4G

After experiencing mobile internet through 3G, there was an increased demand for high-speed internet, which became the primary objective of the fourth generation (4G) communication systems [7]. A further increase in system bandwidth (upto 20 MHz) was needed to support data rate upto 100 Mbps. It is known that the complexity of rake receiver increases exponentially with bandwidth [8] which laid the path for orthogonal frequency division multiplexing (OFDM) to be used in 4G systems. OFDM has a simple transceiver architecture because it uses fast Fourier transform (FFT). In spite of its simplicity, it is resilient to multi-path effects. When Doppler spread is within two percent of sub-carrier bandwidth, the performance of OFDM does not degrade, [9] which makes OFDM a good choice for quasi linear time invariant channel (slowly mobile terminals). OFDM is also known to support frequency domain adaptive modulation and coding (adaptive rate) or frequency domain link adaptation as well as multiple input multiple output (MIMO) antenna–based communication algorithms, which are essential for achieving high spectral efficiency. It has been shown that OFDM is also an optimum waveform for multi-path channel, whose path coefficients do not change with time, i.e., linear time-invariant (LTI) channel. This is because OFDM uses complex exponential as its basis function.

1.5 Fifth Generation (5G) Mobile Communication Systems

5G was envisioned to support applications which along with high data rate also require very low latency and high reliability in high mobility scenarios such as tactile internet [10], autonomous vehicle driving [11], remote vehicle driving etc. In such scenarios, Doppler is also associated with each of the multi-paths of the wireless channel. The wireless channel has delay spread along with Doppler spread which makes it both time and frequency varying in nature. The fifth generation system was also deemed to support machine to machine (M2M) communications and sensor communications to enable various applications like smart home, smart city, accurate weather predictions, etc. The number of such connected devices may rise to 25 billion by 2025, according to a projection by GSMA[12]. These massive number of devices will be distributed anywhere in the world and require a small sporadic data rate in the range of few Kbps in an energy efficient manner. Moreover, 5G is also envisioned to support applications such as virtual reality (VR), augmented reality (AR), ultra-high-definition (UHD) video streaming, high-definition online gaming are examples of applications which require a range of 25–100 Mbps data rate with a small range of 1–4 ms latency.

To conceive above mentioned applications, 3rd Generation Partnership Project (3GPP) has defined three categories of applications which are defined below and also depicted in Figure 1.2.

1. Ultra reliable and low latency communication (URLLC) which requires 0.5 ms latency while maintaining great reliability of 0.99999.
2. Massive machine type communication (MMTC) should support a connection density of one million devices/ KM^2 as well as more than ten years of battery life of MMTC devices.
3. Enhanced mobile broadband (EMBB) provides user experienced data rate of 100 Mbps with 4 ms latency. Additionally, the peak data rate (the best case scenario considering the single user in the system) requirement for EMBB is 20 Gbps in the downlink and 10 Gbps in the uplink.

To elaborate further, Figure 1.3 shows the spider web diagram of different requirements for 5G applications which establishes that the requirements for different applications are diverse.

Need to support higher mobility is one of the several challenges that was enhanced in the requirements of 5G. OFDM is not an optimum choice for high-speed vehicular communication due to its high sensitivity to inter-carrier interference (ICI), which is attributed to Doppler spread. However, the subcarrier bandwidth in OFDM can be increased

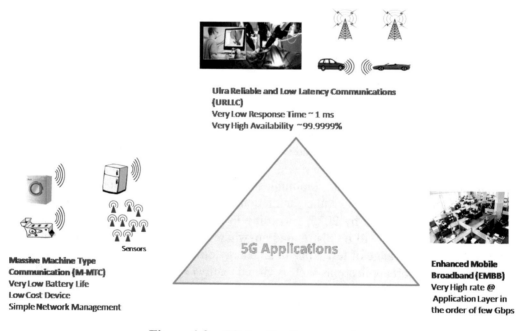

Figure 1.2   5G Application scenarios.

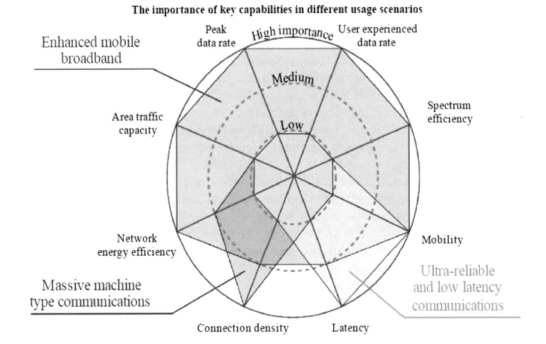

Figure 1.3 Spider web diagram for requirements of 5G applications.

to decrease the ICI at the cost of spectral efficiency loss [9]. A reconfigurable bandwidth framework of OFDM is known as variable sub-carrier bandwidth OFDM (VSB-OFDM) [13, 14, 15, 16], a variant of which is adopted in 5G-NR [17], where VSB reconfigurability is expressed in terms of "numerology". Although sub-carrier bandwidth in 5G-NR can be increased to combat Doppler spread, the provision of proportional decrement of CP length to retain OFDM symbol efficiency introduces interference when both delay and Doppler spreads are significant.

1.6 6G

The last decade saw exabyte of data transmitted over networks, along with the arrival of autonomous vehicles, and widespread use of sensors, IoT, and the beginning of immersive media. Billions of newly connected end points, with varying sensitivity and QoS requires were created. It also exposed several limitations in existing network technologies. In mid 2018, International Telecommunication Union (ITU) started the Network 2030 focus group for exploring the developmet of system technologies for 2030 and beyond. The article [18], is scoped to serve up the communication needs of our society by the year 2030 as well as identify the right set of network technologies required to deliver high-resolution immersive multimedia over the internet, smart IoTs, factory automation, and autonomous vehicles to become a reality.

It mentions, the emergence of Holographic-Type Communications (HTC). It is envisioned that applications will involve local rendering of holograms supported by the ability to transmit and stream holographic data from remote sites, referred to as HTC. There are several useful applications, such as holographic telepresence allowing projection of remote participants in the room, remote troubleshooting and repair applications in difficult to reach areas such as oil drill, space station, etc. HTC systems need to provide a very high bandwidth which involves color depth, resolution, frame rate as well as transmission of volumetric data from multiple viewpoints to account for shifts in tilt, angle, and position of the observer relative to the hologram ("Six Degrees of Freedom").

It also discusses multi-sense networks Applications are in demand that involve optical (video and holograms) and acoustic (audio) senses, touch (tactile), as well as involving the senses of smell and taste. Smell and taste are "near senses" in that their perception involves a direct (chemical) reaction of the agent that is being perceived with a receptor. "Digital lollipops" (http://www.nimesha.info/lollipop.html), are being developed which when inserted into the mouth deliver small currents and differences in temperature to the tongue's papillae (taste sensors) to simulate sensations such as sourness, saltiness, or sweetness. Smell : There are proposals for "transcranial stimulation", that deliver stimuli to areas in the brain responsible for creating sensory sensations. The food industry is expected to reap great benefits from breakthroughs in this area. Assuming that technical challenges that these domains are now facing will be overcome, there will clearly be interesting potential communication applications. Enhanced digital experiences, in particular as smells and tastes that can evoke or amplify emotions will bring major changes in the way the world would transact in the future. Cloud-based medical applications on nutrition etc., would also greatly benefit from these.

Time Engineered Applications: Quick responses and real time experiences have almost become prime factors for smooth functioning in daily life. Especially, industrial automation, autonomous system, and massive networks of sensors where machines interact, the time-factor is very significant. Precision engineering requires that the system must comply to timeliness. Energy efficiency and production facility utilization are extremely important for the manufacturing industry. When waiting time of each piece of equipment in a production chain brought down to almost nil, then utilization is maximized, while energy is saved as failures are minimized. In the industrial internet, small connected entities such as as Programable logic controllers (PLCs), sensors, and actuators require time accuracy of milliseconds and sub-milliseconds. Autonomous traffic systems, are expected to have connected processing points in the order of tens of thousands of vehicles, traffic signals, content, and other components within a small area even of the size of a few kilometers. To enable the smooth operation of such densely inter-connected machinery, the timely delivery of information is necessary. Identical digital environments in applications like online gaming or remote collaborations require true synchronization of multimedia objects at multiple sites. When these are cyber physical systems with the need to move a physical object coordinated in time across sites then time criticality becomes utmost important. Thus, a need is felt to have a communication system that can coordinate between different sources of information such that rendering of such information at multiple sites within the time deadlines is sustainable.

The ITU report [19] and [20] provide detail requirement of such and more applications. There are several articles which discuss the vision of 6G. The evolving nature of the 6G vision limits the detailed discussion of such propositions.

On the technology front related to radio access, some of the recognized dimensions and requirements towards 6G are given below and redrawn in Figure 1.4:

- Migration to larger bandwidth upto 10 GHz.
- Higher frequency bands upto THz.
- Peak data rates in order of few 100s of Gbps.
- Extremely large connection density.
- Nearly 10x increase in realiability.
- 100% improvement in energy efficiency.
- Support for mobility upto 1000 Kmph.

These requirements are challenges of their own. The most important of all things is the radio resource element which carrier information bearing signal from one terminal to the other over the air.

Over the generations of mobile communication systems, the air interface has undergone notable changes which distinguishes one generation from the next. The component of air interface which has evolved of the generation is the radio access technology. In 2G, it was GMSK modulation along with time division mutliplexing. In 3G it was spread spectrum

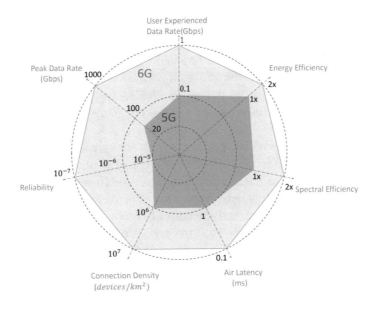

Figure 1.4   Requirements of 6G. [21]

based access technology, while in 4G it was orthogonal time frequency multiplexing. In 5G it was modified version of OFDM, named OFDM numerology where the sub-carrier bandwidth and cyclic prefix interval are made flexible. This was mainly done to tolerate the inter-carrier interference resulting for very high Doppler and phase conditions. Such situation occurs due to extremely high mobility as well as extraordinarily high carrier frequencies. It can be noted from the above that with 6G, spectrum in THz range as well as mobility as high as 1000 kmph are being envisioned to be supported. Such operating scenarios influence the wireless channel heavily which in turn affect the design of wireless transmission technology / radio access technology.

In general, wireless channel is doubly dispersive. A wireless channel has few paths and for each path there is an associated delay and Doppler values. Hence, a wireless channel can be visualized as plurality of scatterers in 2D delay-Doppler domain where each grid point in delay-Doppler domain belongs to a scatterer. Two sub-cases of wireless channels are: (i) linear time invariant channel (LTI) and (ii) linear frequency invariant (LFI) channel. For LTI channel, Doppler values are zero whereas for LFI channel delay values zero. OFDM, used in 4G and in 5G, uses complex exponentials as its bases, is an optimum waveform for LTI channels. OFDM can be termed as delay invariant waveform. Single carriers which use Dirac delta impulses as its bases, is an optimum waveform for LFI channels. Single carrier can be termed as Doppler invariant waveform. Delay invariant waveform performs badly when Doppler effect is also introduced by the channel. In the same way, Doppler invariant waveform performs badly when delays are also introduced by channel. When both delay and Doppler effects are present, a waveform which is delay-Doppler invariant is highly desired. Conventionally, we see the structure of waveform in time-frequency domain. The uncertainty principle constrains finding an appropriate waveform in time-frequency domain. Time-frequency uncertainty principle forbids a waveform which is time-frequency localized to any degree.

The recently proposed waveform Orthogonal Time Frequency Space (OTFS) modulation circumvents this problem by designing waveform in delay-Doppler domain instead. Delay-Doppler domain with unit area is divided into multiple squares. QAM symbols are modulated on these fundamental areas on a delay-Doppler localized pulse. After passing through the channel, these basis pulses are spread at channel grid points within a fundamental area. As a result, pulse energies are spread within a fundamental area and can be despreaded at the receiver. As different QAM symbols are spreaded in their own fundamental area, there is no inter-symbol interference. Thus, such waveform can be considered as delay-Doppler invariant waveform. This effectively implies that equalization can be a simpler task even under presently perceived challenging environment if proper signal synthesis at transmitter and analysis at the receiver is performed. This basic idea was proposed by Prof. Hadani. However, a multicarrier variant of this idea is becoming popular which is based on time-frequency spreading. This variant does not perform pulse shaping in delay-Doppler domain which is the fundamental idea of delay-Doppler invariant waveform. In this book, we explore some of the early developments of OTFS, which can lead to realization transceiver architectures including multiple access techniques.

2

A Summary of Waveforms for Wireless Channels

This chapter presents a literature survey of waveform design in doubly dispersive channels.

2.1 Introduction

The waveform design for doubly dispersive channels has been investigated extensively since 1946. It is vital to look back on these design principles. So, in this chapter, we present a extensive survey of these waveform design principles. This chapter also presents a short summary of legacy waveforms.

2.1.1 Chapter Outline

We begin with mathematical preliminaries such as linear operator on Hilbert space, frames on Hilbert space and Gabor transform in Sec. 2.2. Then, time-frequency analysis tools are presented in Sec. 2.3. The mathematical models of time varying wireless channel are described in Sec. 2.4. Waveform design framework using Gabor transforms and properties of pulse shapes which are relevant for time-varying channel are discussed in Sec. 2.5.

A survey of legacy waveforms such as OFDM (Sec 2.6), windowed OFDM (Sec. 2.8), filtered OFDM (Sec. 2.9), variable sub-carrier OFDM (Sec. 2.7), filter bank multi-carrier (FBMC, Sec. 2.10), universal filtered multi-carrier (UFMC, Sec. 2.11) and generalized frequency division multiplexing (GFDM, Sec. 2.12) is included as well.

2.2 Mathematical Foundation to Time-Frequency Analysis

2.2.1 Hilbert Space

Let V be a complex vector space. An inner product is a mapping $\langle ., . \rangle_V$, from $V \times V \mapsto \mathbb{C}$, such that

1. $\langle \alpha v + \beta w, u \rangle_V = \alpha \langle v, u \rangle_V + \beta \langle w, u \rangle_V, \forall v, w, u \in V$ and $\forall \alpha, \beta \in \mathbb{C}$.
2. $\langle v, w \rangle_V = \overline{\langle w, v \rangle_V}, \forall v, w \in V$.

3. $\langle v, v \rangle_V \geq 0, \forall v \in V.$ If $\langle v, v \rangle_V = 0 \implies v = 0.$

Theorem 2.1 (Cauchy–Swartz inequality) $\forall v, w \in V, |\langle v, w \rangle_V| \leq \langle v, v \rangle_V^{1/2} \langle w, w \rangle_V^{1/2}.$

2.2.2 Norm on Vector Space

A norm on vector space V is a mapping $\|-\| : V \longrightarrow,$ such that

1. $\|v\| \geq 0, \forall v \in V.$
2. $\|v\| = 0 \iff v = 0.$
3. $\|\alpha v\| = |\alpha| \|v\|, \forall \alpha \in \mathbb{C}$ and $\forall v \in V.$
4. $\|v + w\| \leq \|v\| + \|w\|, \forall v, w \in V.$

Lemma 2.1 (norm in Hilbert space) *If V is Hilbert space with inner product $\langle -, - \rangle_V,$ then the norm for the Hilbert space V can be defined as, $\|v\|_V = \langle v, v \rangle_V^{1/2}.$*

2.2.3 Linear Operators on Hilbert Space

Linear mapping, T on a vector space V (i.e., $T : V \to V$) is defined as,

$$T(\alpha v + \beta w) = \alpha T(v) + \beta T(w), \ v, w \in V \ and, \ \beta \in \mathbb{C}.$$

A linear map $T : V \to V$ is bounded if

$$\exists k > 0 : \|Tv\| \leq k\|v\|.$$

The smallest possible of k is called the operator norm on T and is denoted by $\|T\|.$

2.2.3.1 Functional in Hibert Space

Functional is a linear mapping $\Phi : \mathcal{H} \mapsto \mathbb{C}.$ This means all linear mappings in Hilbert space which map to complex numbers are called functionals.

Theorem 2.2 (Reisz Representation Theorem) *Given any functional, $\Phi : \mathcal{H} \mapsto \mathbb{C},$ there exist a $w \in \mathcal{H},$ such that*

$$\Phi(v) = \langle v, w \rangle_{\mathcal{H}} \forall v \in \mathcal{H}. \tag{2.1}$$

Proof. Let, W be a closed null-space of $\Phi,$ such that

$$W = \{v \in \mathcal{H} | \Phi(v) = 0\}.$$

We can choose a vector $u \in W^\perp$ with $\|u\| = 1.$

For any $v \in \mathcal{H},$ consider the vector,

$$z := \Phi(v)u - \Phi(u)v$$

$$\Phi(z) = \Phi(v)\Phi(u) - \Phi(u)\Phi(v) = 0.$$

This shows that $z \in W$. Since, $u \in W^\perp$,

$$\langle z, u \rangle_{\mathscr{H}} = 0$$

$$\implies \langle \Phi(v)u - \Phi(u)v, u \rangle_{\mathscr{H}} = 0$$

$$\implies \Phi(v)\|u\|^2 - \Phi(u)\langle v, u \rangle_{\mathscr{H}} = 0$$

$$\implies \Phi(v) = \langle v, \underbrace{\overline{\Phi(u)}u}_{\in \mathscr{H}} \rangle_{\mathscr{H}}$$

$\square$

2.2.3.2 Adjoint Operator

We consider two Hilbert spaces $\mathscr{H}$ and $\mathscr{K}$. We define a linear operator, $T : \mathscr{K} \mapsto \mathscr{H}$. Corresponding to the operator, T, there exists a adjoint operator, $T^* : \mathscr{H} \mapsto \mathscr{K}$, such that

$$\langle T(v), w \rangle_{\mathscr{H}} = \langle v, T^*(w) \rangle_{\mathscr{K}}, \ \forall v \in \mathscr{K} \text{ and } \forall w \in \mathscr{H}.$$

Properties of Adjoint Operator

Let, $T : \mathscr{K} \mapsto \mathscr{H}$ be a bounded linear operator. Then, the following properties hold

1. $(T^*)^* = T$
2. $\|T\| = \|T^*\|$

Definition 2.1 (Self-adjoint and unitary operators) *Let $\mathscr{H}$ be a Hilbert space and $T : \mathscr{H} \mapsto \mathscr{H}$ be a bounded linear operator.*

1. *The operator T is self-adjoint if $T = T^*$.*
2. *The operator T is unitary if $TT^* = T^*T = \mathscr{I}$*

If operator T is self adjoint, then, $\langle Tv, w \rangle_{\mathscr{H}} = \langle v, T^*w \rangle_{\mathscr{H}} = \langle v, Tw \rangle_{\mathscr{H}}, \forall v, w \in \mathscr{H}$. If T is unitary, then, $\langle Tv, Tw \rangle_{\mathscr{H}} = \langle v, T^*Tw \rangle_{\mathscr{H}} = \langle v, w \rangle_{\mathscr{H}}, \forall v, w \in \mathscr{H}$. When, two vectors, v and w in a Hilbert space are orthogonal, $\langle v, w \rangle_{\mathscr{H}} = 0$. In case T is unitary operator, $\langle Tv, Tw \rangle_{\mathscr{H}} = 0$; that is, a unitary operator preserves orthogonality between two vectors. In addition to that, using the definition of unitary operator, $T^{-1} = T^*$.

Note 1: When operator T is a complex valued matrix. $\langle Tv, w \rangle_{\mathscr{H}} = (Tv)^{\mathrm{H}}w = v^{\mathrm{H}}T^{\mathrm{H}}w = \langle v, T^{\mathrm{H}}w \rangle_{\mathscr{H}}$. This indicates that when T is a matrix, adjoint operator is T^{H}.

2.2.4 Orthonormal Basis for Hilbert Space

A basis $\{u_k\}_{k=1}^{\infty}$ in a Hilbert space $\mathscr{H}$ is an orthonormal basis [1] for $\mathscr{H}$ if $\{u_k\}_{k=1}^{\infty}$ forms an orthonormal system i.e., $\langle u_k, u_{k'} \rangle_{\mathscr{H}} = \delta(k - k')$.

Any vector $v \in \mathscr{H}$ can be expanded in terms of orthonormal basis vectors as,

---

[1] In general, basis can be nonorthogonal which will be discussed later.

$$v = \sum_{k=1}^{\infty} \langle v, u_k \rangle_{\mathcal{H}} u_k, \ \forall v \in \mathcal{H}$$

Theorem 2.3 (Persaval's theorem) *Consider a Hilbert space with $\{u_k\}_{k=1}^{\infty}$ orthonormal basis. Persaval's theorem states that expansion coefficients, $\{\langle v, u_k \rangle_{\mathcal{H}}\}_{k=1}^{\infty} \ \forall v \in \mathcal{H}$ preserves the norm in v , i.e.*

$$\|v\|^2 = \sum_{k=1}^{\infty} |\langle v, u_k \rangle_{\mathcal{H}}|^2, \ \forall v \in \mathcal{H}$$

Proof.

$$\begin{aligned}
\|v\|^2 &= \langle v, v \rangle_{\mathcal{H}} \\
&= \langle \sum_{k=1}^{\infty} \langle v, u_k \rangle_{\mathcal{H}} u_k, \sum_{k'=1}^{\infty} \langle v, u'_k \rangle_{\mathcal{H}} u'_k \rangle_{\mathcal{H}} \\
&= \sum_{k=1}^{\infty} \sum_{k'=1}^{\infty} \langle v, u_k \rangle_{\mathcal{H}} \overline{\langle v, u'_k \rangle_{\mathcal{H}}} \langle u_k, u_{k'} \rangle_{\mathcal{H}} \\
&= \sum_{k=1}^{\infty} \sum_{k'=1}^{\infty} \langle v, u_k \rangle_{\mathcal{H}} \overline{\langle v, u'_k \rangle_{\mathcal{H}}} \delta(k - k') \\
&= \sum_{k=1}^{\infty} |\langle v, u_k \rangle_{\mathcal{H}}|^2
\end{aligned}$$

$\square$

2.2.5 Sequence Space $\mathrm{l}^2(\mathbb{N})$

Sequence space $\mathrm{l}^2(\mathbb{N})$ is a Hilbert space which is defined as,

$$\mathrm{l}^2(\mathbb{N}) := \left\{ \{x_k\}_{k=1}^{\infty} | x_k \in \mathbb{C}, \ \sum_{k=1}^{\infty} |x_k|^2 < \infty \right\}.$$

A vector in $\mathrm{l}^2(\mathbb{N})$ can be denoted as $\mathbf{x} = \{x_k\}_{k=1}^{\infty} = [x_1, \ x_2, \cdots x_n \cdots]$. $\mathrm{l}^2(\mathbb{N})$ is a Hilbert space with respect to the inner product,

$$\langle \mathbf{x}, \mathbf{y} \rangle_{\mathrm{l}^2(\mathbb{N})} = \sum_{k=1}^{\infty} x_k \bar{y}_k, \ \forall \mathbf{x}, \mathbf{y} \in \mathrm{l}^2(\mathbb{N}).$$

Using the definition of inner product, norm in $\mathrm{l}^2(\mathbb{N})$ is defined as,

$$\|\mathbf{x}\| = (\sum_{k=1}^{\infty} |x_k|^2)^{1/2}.$$

2.2.6 Function Spaces

The set consisting of all integrable functions on $\mathbb{R}$ is denoted by $\mathbb{L}^1(\mathbb{R})$, i.e.,

$$\mathbb{L}^1(\mathbb{R}) := \left\{ f : \mathbb{R} \mapsto \mathbb{C} \mid \int_{\infty}^{\infty} |f(x)| \, dx < \infty. \right\}.$$

The norm of $f(x) \in \mathbb{L}^1(\mathbb{R})$ can be given as,

$$\|f\|_1 = \int_{-\infty}^{\infty} |f(x)| \, dx.$$

It should be noted that $\mathbb{L}^1(\mathbb{R})$ is not a Hilbert space, hence, inner product is not defined over $\mathbb{L}^1(\mathbb{R})$.

The set consisting of all square integrable functions on $\mathbb{R}$ is denoted by $\mathbb{L}^2(\mathbb{R})$, i.e.,

$$\mathbb{L}^2(\mathbb{R}) := \left\{ f : \mathbb{R} \mapsto \mathbb{C} \mid \int_{\infty}^{\infty} |f(x)|^2 \, dx < \infty. \right\}.$$

Space $\mathbb{L}^2(\mathbb{R})$ is a Hilbert space equipped with inner product defined as,

$$\langle f, g \rangle_{\mathbb{L}^2(\mathbb{R})} = \int_{\infty}^{\infty} f(x)\overline{g(x)}dx.$$

Norm can be defined as,

$$\|f\|_2 = \left(\int_{\infty}^{\infty} |f(x)|^2 \, dx \right)^{1/2}$$

2.2.7 Fourier Transform

The Fourier transform associates to each function $f \in \mathbb{L}^1(\mathbb{R})$ a new function $\hat{f} : \mathbb{R} \mapsto \mathbb{C}$ given by

$$(\mathscr{F} f)(\gamma) = \hat{f}(\gamma) = \int_{-\infty}^{\infty} f(x)e^{-j2\pi x\gamma}, \; \gamma \in \mathbb{R}.$$

We can extend the Fourier transform to a unitary mapping of $\mathbb{L}^2(\mathbb{R})$ onto $\mathbb{L}^2(\mathbb{R})$. This means that $\langle f, g \rangle_{\mathbb{L}^2(\mathbb{R})} = \langle \hat{f}, \hat{g} \rangle_{\mathbb{L}^2(\mathbb{R})}$ and $\left\| \hat{f} \right\| = \|f\|$.

Assume that $f, \; \hat{f} \in \mathbb{L}^1(\mathbb{R})$, then

$$f(x) = \int_{\infty}^{\infty} \hat{f}(\gamma)e^{j2\pi\gamma x}d\gamma.$$

2.2.7.1 Operators on $\mathbb{L}^2(\mathbb{R})$

For $a \in \mathbb{R}$, the operator $\mathbb{D}_a$ is called translation or shift by a, is defined by,

$$(\mathbb{D}_a f)(x) = f(x - a), \; \forall x \in \mathbb{R} \; and \; f(x) \in \mathbb{L}^2(\mathbb{R}).$$

For $b \in \mathbb{R}$, the operator $\mathbb{M}_b$ is called modulation by b, is defined by,

$$(\mathbb{M}_b f)(x) = e^{j2\pi bx} f(x), \ \forall x \in \mathbb{R} \text{ and } f(x) \in \mathbb{L}^2(\mathbb{R}).$$

The operators $\mathbb{D}_a$ and $\mathbb{M}_b$ are unitary operators of $\mathbb{L}^2(\mathbb{R})$ onto $\mathbb{L}^2(\mathbb{R})$, and the following relation holds

1. $\mathbb{D}_a^{-1} = \mathbb{D}_{-a} = (\mathbb{D}_a)^*$
2. $\mathbb{M}_b^{-1} = \mathbb{M}_{-b} = (\mathbb{M}_b)^*$
3. $(\mathbb{D}_a \mathbb{M}_b f)(x) = e^{-j2\pi ba}(\mathbb{M}_b \mathbb{D}_a f)(x)$
4. $\mathscr{F}\mathbb{D}_a = \mathbb{M}_{-a}\mathscr{F}$.
5. $\mathscr{F}\mathbb{M}_b = \mathbb{D}_b\mathscr{F}$.

2.2.8 Frames in Hilbert Spaces

Definition 2.2 (Frames) *A sequence $\{f_k\}_{k=1}^{\infty}$ of elements in $\mathscr{H}$ is a frame if there exists constants $A, \ B > 0$ such that*

$$A\|f\|^2 \le \sum_{k=1}^{\infty} |\langle f, f_k \rangle_{\mathscr{H}}|^2 \le B\|f\|^2, \ \forall f \in \mathscr{H}$$

The numbers A and B are called frame bounds. The optimal upper frame bound is the infimum over all upper bounds, whereas optimal lower frame bound is supremum over all upper bounds.

2.2.8.1 Frame Operator

If $\{f_k\}_{k=1}^{\infty}$ is a frame for $\mathscr{H}$, then span $\{f_k\}_{k=1}^{\infty} = \mathscr{H}$. The operator $T : \ \mathsf{l}^2(\mathbb{N}) \mapsto \mathscr{H}$, $T\{c_k\}_{k=1}^{\infty} = \sum_{k=1}^{\infty} c_k f_k$ is called the synthesis operator. The adjoint operator is given by, $T^* : \ \mathscr{H} \mapsto \mathsf{l}^2(\mathbb{N})$, $T^* f = \{\langle f, f_k \rangle_{\mathscr{H}}\}_{k=1}^{\infty}$. T^* is also called analysis operator. By composing T and T^*, we obtain frame operator,

$$S : \ \mathscr{H} \mapsto \mathscr{H}, \ Sf = TT^* f = \sum_{k=1}^{\infty} \langle f, f_k \rangle_{\mathscr{H}} f_k.$$

$$\langle Sf, f \rangle = \langle \sum_{k=1}^{\infty} \langle f, f_k \rangle f_k, f \rangle = \sum_{k=1}^{\infty} |\langle f, f_k \rangle|^2$$

Frame operator is bounded, invertible, self-adjoint, and positive. If $\{f_k\}_{k=1}^{\infty}$ is a frame with frame operator S. Then,

$$f = \sum_{k=1}^{\infty} \langle f, S^{-1} f_k \rangle_{\mathscr{H}} f_k, \ \forall f \in \mathscr{H},$$

where, the sequence $\{S^{-1} f_k\}_{k=1}^{\infty}$ is called canonical dual frame corresponding to frame $\{f_k\}_{k=1}^{\infty}$. If $\{f_k\}_{k=1}^{\infty}$ is a frame for $\mathscr{H}$ with bounds A and B, $\{S^{-1} f_k\}_{k=1}^{\infty}$ is also a frame for $\mathscr{H}$ with bounds B^{-1} and A^{-1}.

Let $\{f_k\}_{k=1}^{\infty}$ be a frame with synthesis operator T and frame operator S. Then, $T^{\dagger}f = \{\langle f, S^{-1}f_k \rangle_{\mathscr{H}}\}_{k=1}^{\infty}$, where $T^{\dagger}$ is pusedo inverse operator of T.

The optimal frame bounds $A = \left\|S^{-1}\right\|^{-1} = \left\|T^{\dagger}\right\|^{2}$ and $B = \|S\| = \|T\|^{2}$.

If $\{f_k\}_{k=1}^{\infty}$ is a frame for $\mathscr{H}$ with bounds A and B. Let, $U : \mathscr{H} \mapsto \mathscr{H}$ be a bounded linear operator. Then, $\{Uf_k\}_{k=1}^{\infty}$ is a frame sequence with frame bounds $A\left\|U^{\dagger}\right\|^{-2}$ and $B\|U\|^{2}$. As a special case when U is a unitary operator, $\{Uf_k\}_{k=1}^{\infty}$ is a frame sequence with frame bounds A and B.

Let us compose T^* and T, we obtain $T^*T : l^2(\mathbb{N}) \mapsto l^2(\mathbb{N}), T^*T\{c_k\}_{k=1}^{\infty} = \{\langle \sum_{l=1}^{\infty} c_l f_l, f_k \rangle_{\mathscr{H}}\}_{k=1}^{\infty}$. We can write above mapping as a matrix vector multiplication as, $\boldsymbol{G}\mathbf{c}$, where, $\boldsymbol{G}$ is called Gram matrix, whose $(j, k)^{\text{th}}$ element can be given as, $\{\langle f_k, f_j \rangle_{\mathscr{H}}\}$, $j, k \in \mathbb{N}$.

2.2.8.2 Reisz Basis

A Reisz basis for $\mathscr{H}$ is a family of the form $\{Uu_k\}_{k=1}^{\infty}$, where $\{u_k\}_{k=1}^{\infty}$ is an orthonormal basis for $\mathscr{H}$ and $U : \mathscr{H} \mapsto \mathscr{H}$ is a bounded bijective (one to one and onto) operator. A Reisz basis $\{f_k\}_{k=1}^{\infty}$ for $\mathscr{H}$ is a frame for $\mathscr{H}$, and Riesz basis bounds coincide with the frame bounds. The dual Reisz basis is $\{S^{-1}f_k\}_{k=1}^{\infty}$. For a Reisz basis, frames and dual frames are bidelete hyphenorthogonal i.e., $\langle f_k, S^{-1}f_l \rangle_{\mathscr{H}} = \delta(k-l)$.

A frame which is not a Reisz basis is said to be overcomplete. For a frame which is not a Reisz basis, there exists coefficients, $\{c_k\}_{k\in\mathbb{N}} \in l^2(\mathbb{N})/\{0\}\}$ for which, $\sum_{k=1}^{\infty} c_k f_k = 0$.

Definition 2.3 (Bessel sequence) *A sequence $\{f_k\}_{k=1}^{\infty}$ is called a Bessel sequence if*

$$\sum_{k=1}^{\infty} |\langle f, f_k \rangle_{\mathscr{H}}|^2 \leq B\|f\|^2, \ \forall f \in \mathscr{H}$$

The bound of a Bessel sequence is the smallest B that satisfies the corresponding inequality.

2.2.8.3 Tight Frame

Frame $\{f_k\}_{k=1}^{\infty}$ is a tight frame if lower and upper frame bounds are equal i.e., $A = B$. Thus, $\sum_{k=1}^{\infty} |\langle f, f_k \rangle_{\mathscr{H}}|^2 = A\|f\|^2 \implies \langle Sf, f \rangle = \langle Af, f \rangle \implies S = A\mathscr{I}$. Thus, $f = \frac{1}{A}\sum_{k=1}^{\infty} \langle f, f_k \rangle_{\mathscr{H}} f_k$.

If $\{f_k\}_{k=1}^{\infty}$ is a frame for $\mathscr{H}$ with frame operator S. Then, $\{S^{-1/2}f_k\}_{k=1}^{\infty}$ is a tight frame with frame bound equal to one. i.e., $f = \sum_{k=1}^{\infty} \langle f, S^{-1/2}f_k \rangle_{\mathscr{H}} S^{-1/2}f_k$.

2.2.8.4 Dual Frame

Let $\{f_k\}_{k=1}^{\infty}$ be a frame for $\mathscr{H}$. The dual frames for $\{f_k\}_{k=1}^{\infty}$ are used to obtain f from $\{c_k\}_{k\in\mathbb{N}} = \{\langle f, f_k \rangle_{\mathscr{H}}\}$ as $f = \sum_{k=1}^{\infty} c_k g_k$, where , is the dual frame for $\{f_k\}_{k=1}^{\infty}$. When frames are exact, dual frames are unique but when frames are over-complete dual frames are not unique.

Let $\{f_k\}_{k=1}^{\infty}$ be a frame for $\mathcal{H}$. The dual frames of $\{f_k\}_{k=1}^{\infty}$ are precisely the family

$$= \{S^{-1}f_k + h_k - \sum_{j=1}^{\infty} \langle S^{-1}f_k, f_j \rangle_{\mathcal{H}} h_j\}_{k=1}^{\infty},$$

where, $\{h_k\}_{k=1}^{\infty}$ is a bessel sequence in $\mathcal{H}$. When frame is a Reisz basis $\langle f_k, S^{-1}f_l \rangle_{\mathcal{H}} = \delta(k-l)$, so, $\langle S^{-1}f_k, f_j \rangle_{\mathcal{H}} h_j = h_k$. Thus, dual Reisz basis is unique and given by $= \{S^{-1}f_k\}_{k=1}^{\infty}$.

2.2.9 Gabor Transform

$\{g_{m,k}\}_{m,k\in\mathbb{N}} = \{g(t-mT)e^{j2\pi Ft}\}_{m,k\in\mathbb{N}}$ constitutes a Gabor frame if and only if $TF < 1$. If $\{g_{m,k}\}_{m,k\in\mathbb{N}}$ is a Gabor frame with frame bounds A and B, then

$$FA \le \sum_{m\in\mathbb{N}} |g(t-mT)|^2 \le FB.$$

Suppose $\{g_{m,k}\}_{m,k\in\mathbb{N}}$ is a Gabor frame for $\mathbb{L}^2(\mathbb{R})$, The operator $\mathscr{T} : l^2(\mathbb{N}) \mapsto \mathbb{L}^2(\mathbb{R})$, $\mathscr{T}\{c_{m,k}\}_{m,k\in\mathbb{N}} = \sum_{m,k\in\mathbb{N}} c_{m,k}g_{m,k}$ is called the synthesis operator. The adjoint operator is given by, $T^* : \mathcal{H} \mapsto l^2(\mathbb{N})$, $\mathscr{T}^*f = \{\langle f, g_{m,k} \rangle_{\mathbb{L}^2(\mathbb{R})}\}_{m,k\in\mathbb{N}}$. $\mathscr{T}^*$ is also called analysis operator. By composing T and T^*, we obtain frame operator,

$$S : \mathbb{L}^2(\mathbb{R}) \mapsto \mathbb{L}^2(\mathbb{R}), \ Sf = \mathscr{T}\mathscr{T}^*f = \sum_{m,k\in\mathbb{N}} \langle f, g_{m,k} \rangle_{\mathbb{L}^2(\mathbb{R})} g_{m,k}.$$

$$\langle Sf, f \rangle = \langle \sum_{k=1}^{\infty} \langle f, f_k \rangle f_k, f \rangle = \sum_{k=1}^{\infty} |\langle f, f_k \rangle|^2$$

Theorem 2.4 (Ron–Shen duality principle) *Suppose Let $g \in \mathbb{L}^2(\mathbb{R})$ and $T, F > 0$ be given. Then, the Gabor system $\{g_{k,m}\}_{k,m\in\mathbb{N}} = \{\mathbb{M}_{kF}\mathbb{D}_{mT}g\}_{k,m\in\mathbb{N}}$ is a frame for $\mathbb{L}^2(\mathbb{R})$ with bounds A and B if and only if $\{\mathbb{M}_{\frac{k}{T}}\mathbb{D}_{\frac{m}{F}}g\}_{k,m\in\mathbb{N}}$ is a Reisz sequence with bounds TFA and TFB.*

Theorem 2.5 (Dual frame) *Let $g \in \mathbb{L}^2(\mathbb{R})$ and $T, F > 0$ be given and assume $\{g_{k,m}\}_{k,m\in\mathbb{N}}$ be a frame for $\mathbb{L}^2(\mathbb{R})$. Then, for any $f \in \mathbb{L}^2(\mathbb{R})$ for which $\{\mathbb{M}_{kF}\mathbb{D}_{mT}f\}_{m,n\in\mathbb{Z}}$ is a Bessel sequence, the function*

$$h = S^{-1}g + f - \sum_{m,n\in\mathbb{Z}} \langle S^{-1}g, \mathbb{M}_{kF}\mathbb{D}_{mT}g \rangle_{\mathbb{L}^2(\mathbb{R})} \mathbb{M}_{kF}\mathbb{D}_{mT}f$$

generate a dual frame $\{\mathbb{M}_{kF}\mathbb{D}_{mT}h\}_{k,m\in\mathbb{Z}}$ of $\{\mathbb{M}_{kF}\mathbb{D}_{mT}g\}_{m,n\in\mathbb{Z}}$.

Theorem 2.6 (dual lattice) *Let $g, h \in \mathbb{L}^2(\mathbb{R})$ and $T, F > 0$ be given. Then, if the two Gabor systems $\{\mathbb{M}_{kF}\mathbb{D}_{mT}g\}_{k,m\in\mathbb{N}}$ and $\{\mathbb{M}_{kF}\mathbb{D}_{mT}\gamma\}_{k,m\in\mathbb{N}}$ are dual frames if and only if,*

$$\langle g, \mathbb{M}_{\frac{k}{T}}\mathbb{D}_{\frac{m}{F}}\gamma \rangle_{\mathbb{L}^2(\mathbb{R})} = TF\delta(k)\delta(m).$$

For a Gabor frame, transmitted signal, $s(t) = \sum_{m,k} d_{m,n} g_{m,k}(t)$. Let us first discuss, the signal recovery without wireless channel and noise. At the receiver, estimated data can be given as,

$$\hat{d}_{m,k} = \langle s, \mathrm{M}_{\frac{k}{T}} \mathrm{D}_{\frac{m}{F}} \gamma \rangle_{\mathrm{L}^2(\mathbb{R})}.$$

When, $TF < 1$, $\hat{d}_{m,k}$'s are nonunique. Hence, the estimated data can be written as,

$$\hat{d}_{m,k} = d_{m,k} + e, \tag{2.2}$$

where, $e \in N_{\mathscr{T}}$, where $\mathscr{T}$ is synthesis operator. This problem pose a theoretical bottleneck to analyze FTN system in Gabor setting.

Let us consider a system with N number of frequency slots and M number of time slots. Transmitted signal can be given as,

$$x(t) = \sum_{m=0}^{M-1} \sum_{k=0}^{N-1} g(t - mT) e^{j2\pi kFt} = \sum_{m=0}^{M-1} \sum_{k=0}^{N-1} g_{k,m}(t).$$

Theorem 2.7 *When, $TF = 1$, $\{g_{k,m}\}_{k \in \mathbb{Z}_N, m \in \mathbb{Z}_M}$ forms a Reisz basis for a subspace $A \subset \mathrm{L}^2(\mathbb{R})$. When, $TF < 1$, $\{g_{k,m}\}_{k \in \mathbb{Z}_{N'}, m \in \mathbb{Z}_{M'}}$ forms frame for $A \subset \mathrm{L}^2(\mathbb{R})$ where $M'N' > MN$.*

2.3 Time–Frequency Foundations

2.3.1 Time–Frequency Uncertainty Principle

If $f(x) \in \mathrm{L}^2(\mathbb{R})$ and $a, b \in \mathbb{R}$ are arbitrary, then

$$\left(\int_{\mathbb{R}} (x - a)^2 |f(x)|^2 \, dx \right)^{1/2} \left(\int_{\mathbb{R}} (\gamma - b)^2 \left| \hat{f}(\gamma) \right|^2 \, d\gamma \right)^{1/2} \geq \frac{1}{4\pi} \|f\|_2^2.$$

Equality holds if and only if $f(x)$ is a multiple of $\mathbb{D}_a \mathbb{M}_b \psi_c(x) = e^{j2\pi b(x-a)} e^{-\pi(x-a)^2/c}$ for some $a, b \in \mathbb{R}$ and $c > 0$, where, $\psi_c(x) = e^{-\pi x^2/c}$ is a Gaussian function. The uncertainty principle loosely formulated as, "A realizable signal occupies a region of area at least one in the time-frequency plane."

2.3.2 Short Time Fourier Transform

Fix a function $g(t) \neq 0$ (called the window function). Then, the short-time Fourier transform (STFT) of a function $f(t)$ with respect to $g(t)$ is defined as,

$$V_g f(\tau, \gamma) = \langle f, \mathbb{M}_\gamma \mathbb{D}_\tau g \rangle_{\mathrm{L}^2(\mathbb{R})} = \int_{\mathbb{R}} f(t) \overline{g(t - \tau)} e^{-j2\pi t\gamma} dt, \; for \; \tau, \gamma, t \in \mathbb{R}.$$

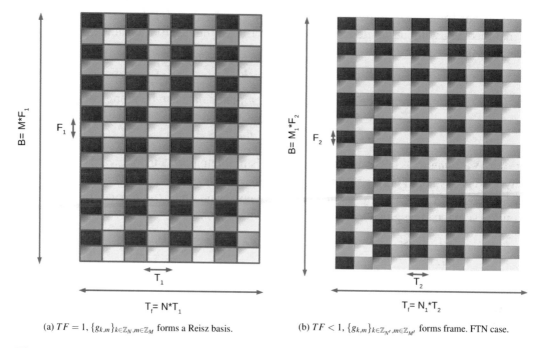

(a) $TF = 1$, $\{g_{k,m}\}_{k\in\mathbb{Z}_N, m\in\mathbb{Z}_M}$ forms a Reisz basis.   (b) $TF < 1$, $\{g_{k,m}\}_{k\in\mathbb{Z}_{N'}, m\in\mathbb{Z}_{M'}}$ forms frame. FTN case.

Figure 2.1   This figure shows TF space for $TF = 1$ and $TF < 1$ case. In both cases, total time duration T_f and bandwidth B are same. Each square represents a basic pulse shape.

2.3.2.1 Properties

1. $V_g f(\tau, \gamma) = e^{-j2\pi\tau\gamma} V_{\hat{g}} \hat{f}(\gamma, -\tau)$.
2. $V_g \mathbb{D}_a \mathbb{M}_b f(\tau, \gamma) = e^{-j2\pi a\gamma} V_g f(\tau - a, \gamma - b)$.
3. $\langle V_{g_1} f_1, V_{g_2} f_2 \rangle_{\mathbb{L}^2(\mathbb{R}^2)} = \langle f_1, f_2 \rangle_{\mathbb{L}^2(\mathbb{R})} \overline{\langle g_1, g_2 \rangle_{\mathbb{L}^2(\mathbb{R})}}$.
4. If $\|g\|_2 = 1$ then, $\|f\|_2 = \|V_g f\|_2$. So, in this case STFT is an isometry from $\mathbb{L}^2(\mathbb{R})$ to $\mathbb{L}^2(\mathbb{R}^2)$.
5. Suppose that g and $h \in \mathbb{L}^2(\mathbb{R})$ and $\langle g, h \rangle \neq 0$. Then, for all $f \in \mathbb{L}^2(\mathbb{R})$,

$$f = \frac{1}{\langle f, h \rangle} \iint_{\mathbb{R}^2} V_g f(\tau, \gamma) \mathbb{M}_\gamma \mathbb{D}_\tau h \, d\tau \, d\gamma$$

2.3.3 Ambiguity Function

The ambiguity function of $f \in \mathbb{L}^2(\mathbb{R})$ is defined to be

$$\mathscr{A}_f(\tau, \gamma) = \int_{\mathbb{R}} f(t + \frac{\tau}{2}) \overline{f(t - \frac{\tau}{2})} e^{-j2\pi t\gamma} dt = e^{j2\pi\tau\gamma} V_f f(\tau, \gamma).$$

The cross ambiguity function of f and $g \in \mathbb{L}^2(\mathbb{R})$ is

$$\mathscr{A}_{f,g}(\tau, \gamma) = \int_{\mathbb{R}} f(t + \frac{\tau}{2}) \overline{g(t - \frac{\tau}{2})} e^{-j2\pi t\gamma} dt = e^{j2\pi\tau\gamma} V_g f(\tau, \gamma).$$

Most properties of the STFT carry over to the ambiguity function. The most notable difference between STFT and ambiguity function occurs in the inversion formula. $\mathscr{A}_f$ determines f only to a phase factor. If, $|c| = 1$, then $\mathscr{A}_{cf} = \mathscr{A}_f$. $f(t)$ upto a phase factor can be obtained as,

$$ f(t) = \frac{1}{f(0)} \int_{\mathbb{R}} \mathscr{A}_f(t, \gamma) e^{j\pi t \gamma} d\gamma, $$

where, all other solutions are cf, where $|c| = 1$.

2.4 Linear Time Varying Channel

Wireless channel, in general, is linear time-varying (LTV). When a signal goes through a typical wireless channel it gets relected , refracted, and defracted from multiple scatterers. This multi-path effect results in the time dispersion of the signal. Additionally, these refractors, as well as transceivers, can be mobile which results in frequency dispersion of the signal. The channel effects can be modeled as a linear transversal filter. However, when there is a relative motion between the incoming wavefront and the receiver terminal then the channel no longer remains static. The coefficients of the channel impulse response change with time, thus, resulting in a linear time varying channel. Here, an LTV channel is denoted as a linear operator $\mathscr{H}$ which acts on the transmitted signal $x(t)$ and yields the received signal $y(t) = (\mathscr{H}x)(t)$. In the next subsections, we present different descriptions of $\mathscr{H}$.

2.4.1 Delay-Doppler Spreading Function ($\mathscr{S}_{\mathscr{H}}(\tau, \nu)$)

Delay-Doppler spreading function $\mathscr{S}_{\mathscr{H}}(\tau, \nu)$ represents overall complex attenuation for all the paths associated with a particular delay-Doppler tuple value (τ, ν). It describes how the delayed and Doppler shifted version $x(t - \tau) e^{j2\pi\nu t}$ of transmit signal $x(t)$ contributes to the received signal $r(t)$. In other words, $\mathscr{S}_{\mathscr{H}}(\tau, \nu)$ describes the spreading of transmit signal in time and frequency. If τ_{max} and ν_{max} are maximum delay and Doppler spread of the channel, then the received signal can be written as,

$$ y(t) = \int_{-\nu_{max}}^{\nu_{max}} \int_{0}^{\tau_{max}} \mathscr{S}_{\mathscr{H}}(\tau, \nu) x(t - \tau) e^{j2\pi\nu t} d\tau d\nu \tag{2.3} $$

A dual representation of the channel's TF dispersion in frequency is given as,

$$ Y(f) = \int_{-\nu_{max}}^{\nu_{max}} \int_{0}^{\tau_{max}} \mathscr{S}_{\mathscr{H}}(\tau, \nu) X(f - \nu) e^{-j2\pi\tau(f-\nu)} d\tau d\nu \tag{2.4} $$

We can also obtain channel's TF dispersion in a joint TF domain. Taking STFT of (2.3), we can obtain,

$$ V_g y(t, f) = \int_{-\nu_{max}}^{\nu_{max}} \int_{0}^{\tau_{max}} \mathscr{S}_{\mathscr{H}}(\tau, \nu) V_g x(t - \tau, f - \nu) e^{-j2\pi\tau(f-\nu)} d\tau d\nu. \tag{2.5} $$

This equation demonstrates that signal in spread in joint TF domain by the channel. If signal is bounded in TF domain i.e., $V_g x(t, f) = 0$ for some $t > T'$ and $|f| > F'$, then received signal is also bounded i.e., $V_g y(t, f) = 0$ for $t > T' + \tau_{max}$ and $|f| > F' + \nu_{max}$.

2.4.2 Time-Varying Transfer Function ($\mathscr{L}_{\mathscr{H}}(t, f)$)

Time dispersiveness of the channel corresponds to frequency selectivity, whereas frequency dispersiveness corresponds to time selectivity. Joint TF selectivity is described by time-varying transfer function ($\mathscr{L}_{\mathscr{H}}(t, f)$)), which can be given as,

$$\mathscr{L}_{\mathscr{H}}(t, f) = \int_{-\nu_{max}}^{\nu_{max}} \int_{0}^{\tau_{max}} \mathscr{S}_{\mathscr{H}}(\tau, \nu) e^{j2\pi\nu t - f\tau} d\tau d\nu. \tag{2.6}$$

Received signal can be given as [2],

$$y(t) = \int_{-\infty}^{\infty} \mathscr{L}_{\mathscr{H}}(t, f) S(f) e^{j2\pi t f} df \tag{2.7}$$

2.4.3 Time-Varying Impulse Response ($h(t, \tau)$)

Wireless channel can be represented as a continuous tapped delay line. When a delay τ is fixed, $h(t, \tau)$ is a function of $'t'$ describing time-varying tap weight. For a fixed $'t'$, $h(t, \tau)$ is a function of τ and represents weights of delay τ. Suppose there are L number of multi-paths, $h(t, \tau)$ can be given as,

$$h(t, \tau) = \sum_{l=0}^{L} w_i(t) \delta(t - \tau_i), \tag{2.8}$$

where, $w_i(t)$ and τ_i are the complex weight and delay for i^{th} path. Received signal, $y(t)$, can be obtained as,

$$r(t) = \int_{0}^{\tau_{max}} h(t, \tau) x(t - \tau) d\tau. \tag{2.9}$$

We can obtain $h(t, \tau)$ from $\mathscr{S}_{\mathscr{H}}(\tau, \nu)$ by taking IFFT along Doppler variable ν and can be given as,

$$h(t, \tau) = \int_{-\nu_{max}}^{\nu_{max}} \mathscr{S}_{\mathscr{H}}(\tau, \nu) e^{j2\pi t \nu} d\nu. \tag{2.10}$$

Similarly, $\mathscr{L}_{\mathscr{H}}(t, f)$ can be obtained from $h(t, \tau)$ by taking FFT along τ variable and can be given as,

$$\mathscr{L}_{\mathscr{H}}(t, f) = \int_{0}^{\tau_{max}} h(t, \tau) e^{-j2\pi f\tau} d\tau. \tag{2.11}$$

Figure 2.2 describes conversion between different channel representation.

2.4.4 Linear Time Invariant (LTI) Channel

We will take a little detour to discuss LTI channel as a special case of LTV channel. Here, we assume the channel to be LTI. For LTI system $\mathbb{D}_\tau$ commutes with $\mathscr{H}$ i.e., $\mathbb{D}_\tau \mathscr{H} = \mathscr{H} \mathbb{D}_\tau$.

---

[2]It should be noted that (2.7) is not a simple inverse Fourier transform as $\mathscr{L}_{\mathscr{H}}(t, f)$ also depends on t.

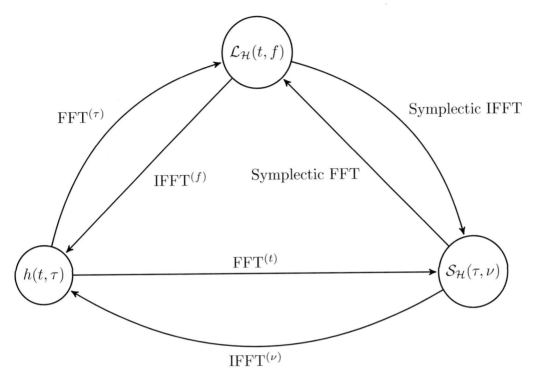

Figure 2.2 Conversions between different channel representations. FFT$^{(-)}$/ IFFT$^{(-)}$ represents FFT/ IFFT taken for $'-'$ variable.

When $\nu_{max} = 0$, LTV channel becomes LTI. Thus, $\mathscr{S}_{\mathcal{H}}(\tau, \nu) = h(\tau)\delta(\nu)$, and (2.3) converges to channel convolution equation and can be given as,

$$y(t) = \int_0^{\tau_{max}} h(\tau)x(t - \tau)d\tau. \tag{2.12}$$

Similarly, $h(t, \tau) = h(\tau)$ and (2.9) also converges to (2.12). Now using (2.11) and $h(t, \tau) = h(\tau)$, $\mathscr{L}_{\mathcal{H}}(t, f) = H(f)$ where $H(f) = \mathscr{F}h(\tau)$ is channel's frequency response. Received signal in frequency domain can be given as,

$$Y(f) = H(f)X(f). \tag{2.13}$$

2.4.5 Stochastic Description

We consider reflectivity of any two distinct scatterers are uncorrelated. This implies that the spreading function $\mathscr{S}_{\mathcal{H}}(\tau, \nu)$ is a 2D white (but non stationary) process i.e.,

$$E(\mathscr{S}_{\mathcal{H}}(\tau, \nu)\mathscr{S}_{\mathcal{H}}(\tau', \nu')) = \mathscr{C}_{\mathcal{H}}(\tau, \nu)\delta(\tau - \tau')\delta(\nu - \nu'), \tag{2.14}$$

where, $\mathscr{C}_{\mathcal{H}}(\tau, \nu) \geq 0$ is channel's scattering function. Mean delay and mean Doppler shift describe average location of scattering function in the (τ, ν) plane. Spread of

$\mathscr{C}_{\mathscr{H}}(\tau, \nu)$ about $(\bar{\tau}, \bar{\nu})$ can be measured by delay spread and Doppler spread which can be computed as,

$$\sigma_\tau = \frac{1}{\rho_{\mathscr{H}}} \int_{-\infty}^{\infty} \int_{-\infty}^{\infty} (\tau - \bar{\tau})^2 \mathscr{C}_{\mathscr{H}}(\tau, \nu) d_\tau d_\nu \tag{2.15}$$

$$\sigma_\nu = \frac{1}{\rho_{\mathscr{H}}} \int_{-\infty}^{\infty} \int_{-\infty}^{\infty} (\nu - \bar{\nu})^2 \mathscr{C}_{\mathscr{H}}(\tau, \nu) d_\tau d_\nu, \tag{2.16}$$

where $\rho_{\mathscr{H}}$ is the path-loss exponent which is the volume of $\mathscr{C}_{\mathscr{H}}(\tau, \nu)$ i.e., $\rho_{\mathscr{H}} = \int_{-\nu_{max}}^{\nu_{max}} \int_{0}^{\tau_{max}} \mathscr{S}_{\mathscr{H}}(\tau, \nu) d\tau d\nu$.

WSS channels are stationary in time, whereas uncorrelation among scatterers brings stationarity in frequency. Which means that statistics of a WSSUS channel does not change with time and frequency. Thus, TF transfer function $\mathscr{L}_{\mathscr{H}}(t, f)$ is 2D stationary i.e.,

$$E[\mathscr{L}_{\mathscr{H}}(t, f)\mathscr{L}_{\mathscr{H}}(t', f')] = R_{\mathscr{H}}(t - t', f - f') = R_{\mathscr{H}}(\Delta t, \Delta f), \tag{2.17}$$

where $R_{\mathscr{H}}(\Delta t, \Delta f)$ is channel's TF correlation function. Time correlation and frequency correlation of channel can be obtained as a marginal of $R_{\mathscr{H}}(\Delta t, \Delta f)$ i.e. $r_{\mathscr{H}}(\Delta t) = R_{\mathscr{H}}(\Delta t, 0)$ and $r_{\mathscr{H}}(\Delta f) = R_{\mathscr{H}}(0, \Delta f)$.

2.4.6 Under-Spread Property of Wireless Channel

Generally, the spread of scattering function $\mathscr{C}_{\mathscr{H}}(\tau, \nu)$ has compact support. Suppose, τ_{max} and ν_{max} are the channels' maximum delay and Doppler, respectively. Compact support of the scattering function can be defined as the smallest rectangle having length of $2\tau_{max}$ and breadth of $2\nu_{max}$. Area of such rectangle is $a_{\mathscr{H}} = 4\tau_{max}\nu_{max}$. A WSSUS channel is said to be under-spread if $a_{\mathscr{H}} < 1$ [22] [23]. Real world wireless channels are always under-spread.

Another under-spread description is developed by Matz in [24] [25]. The notation of an effective support of $\mathscr{C}_{\mathscr{H}}(\tau, \nu)$ is described as $\sigma_{\mathscr{H}} = \sigma_\tau \sigma_\nu$. The dispersion-under-spread expresses the fact that channel's dispersion spread is small. A channel is said to be dispersion-under-spread if $\sigma_{\mathscr{H}} < 1$.

2.4.7 Physical Discrete Path Model

We consider a wireless channels having N_P number of distinct paths. For each path $p = \mathbb{N}[1\ P]$, there is an associated delay, τ_p and Doppler ν_p. In the physical discrete path model $\mathscr{L}_{\mathscr{H}}(t, f)$ can be given as,

$$\mathscr{L}_{\mathscr{H}}(t, f) = \sum_{p=1}^{N_P} c_p e^{-j2\pi\tau_p f} e^{j2\pi\nu_p t}, \tag{2.18}$$

where, $c_p = a_p e^{j\phi_p}$ is complex attenuation corresponding to p^{th} path. Over the small time-scales of the interest, we assume that N_p and $\{a_p,\ \tau_p,\ \nu_p\}$ are deterministic but unknown, whereas, ϕ_p is random variable that is uniformly distributed over $[-\pi \ \pi]$ and is independent for different paths. Consider a data packet of length T_f seconds and bandwidth of B hertz. Thus, wireless channel $\mathscr{L}_{\mathscr{H}}(t, f)$ is restricted to T_f second and B Hz. Scattering function, $\mathscr{S}_{\mathscr{H}}(\tau, \nu)$ can be obtained as,

$$\mathscr{S}_{\mathscr{H}}(\tau, \nu) = \frac{1}{BT_f} \int_0^{T_f} \int_{-B/2}^{B/2} \mathscr{L}_{\mathscr{H}}(t, f)e^{-j2\pi\nu t}e^{j2\pi\tau f} dt df \quad (2.19)$$

$$= \frac{1}{BT_f} \int_0^{T_f} \int_{-B/2}^{B/2} \sum_{p=1}^{N_P} c_p e^{-j2\pi t[\nu-\nu_p]} e^{j2\pi f[\nu-\nu_p]} dt df \quad (2.20)$$

$$= \sum_{p=1}^{P} \underbrace{c_n e^{j2\pi T_f[\nu-\nu_p]}}_{\tilde{c}_n} Sinc(T_f[\nu - \nu_p])Sinc(B[\tau - \tau_p]), \quad (2.21)$$

where, $Sinc(\vartheta) = \frac{\sin(\pi\vartheta)}{\pi\vartheta}$ is a sinc function. In the case when wireless channel is not restricted in TF domain, $\lim_{T_f \to \infty} Sinc(T_f[\nu - \nu_p]) = \delta(\nu - \nu_p)$ and $\lim_{T_f \to \infty} Sinc(B[\tau - \tau_p]) = \delta(\tau - \tau_p)$; and, thus, $\mathscr{S}_{\mathscr{H}}(\tau, \nu) = \sum_{p=1}^{P} \tilde{c}_n \delta(\nu - \nu_p)\delta(\tau - \tau_p)$.

2.4.7.1 Virtual Channel Representation: Sampling in Delay-Doppler Domain

The exact values of physical delay and Doppler shifts are not critical from a communication perspective. It is resolvable delays and Doppler shifts which are important from communication perspective. Suppose a communication system has total system bandwidth of B Hz and total frame duration of T_f seconds. Moreover, $M = \lceil B\tau_{max} \rceil$ and $N = \lceil T_f\nu_{max} \rceil$. To get virtual channel representation, delay axis is sampled with sampling distance $\Delta\tau = \frac{1}{B}$ seconds, whereas Doppler axis is sampled with sampling distance of $\Delta\nu = \frac{1}{T_f}$ Hz. Thus, delay-Doppler plane is divided into multiple boxes of area $\Delta\tau \times \Delta\nu$. Physical channels which lie within a boxes are quantized to nearest virtual channel delay-Doppler values. Figure 2.3 illustrates such virtual channel representation.

Let $S_{\tau,l}$, $l \in \mathbb{N}[0 \ M - 1]$, collects all the physical paths which have delays between $\frac{l}{B} - \frac{1}{2B}$ seconds and $\frac{l}{B} + \frac{1}{2B}$ seconds, i.e.,

$$S_{\tau,l} = \left\{ p : \frac{l}{B} - \frac{1}{2B} < \tau_p \le \frac{l}{B} + \frac{1}{2B} \right\}. \quad (2.22)$$

Similarly, $S_{\nu,m}$, $m \in \mathbb{N}[-(N-1) \ N - 1]$, collects all the physical paths having Doppler values between $\frac{m}{T_f} - \frac{1}{2T_f}$ and $\frac{m}{T_f} + \frac{1}{2T_f}$, i.e.,

$$S_{\nu,m} = \left\{ p : \frac{m}{T_f} - \frac{1}{2T_f} < \nu_p \le \frac{m}{T_f} + \frac{1}{2T_f} \right\}. \quad (2.23)$$

After sampling delay and Doppler domain in (2.21), channel spreading function for virtual channel can be given as,

$$\mathscr{S}_{\mathscr{H}}(l, m) \approx \sum_{p \in S_{\tau,l}} \tilde{c}_p, \quad (2.24)$$

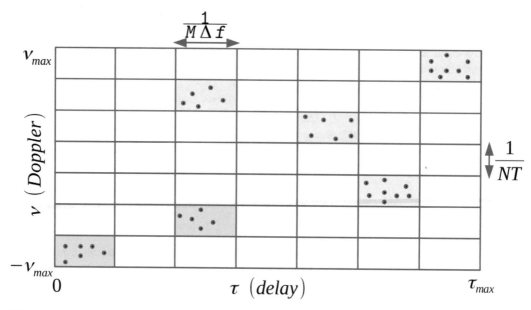

Figure 2.3   Illustration of the virtual channel representation and path partitioning in delay-Doppler domain. Dots in the figure represent real physical delay and Doppler values. Corner points of each box are quantized delay and Doppler values.

where, phase and attenuation due to the Sinc functions are incorporated into c_n's in the approximation. Thus, the received signal can be approximated as,

$$r(t) \approx \sum_{l=0}^{M-1} \sum_{m=-(N-1)}^{N-1} \mathscr{S}_{\mathscr{H}}(l,m)x\left(t - \frac{l}{B}\right)e^{j2\pi\frac{m}{T_f}t}. \tag{2.25}$$

2.5 Waveform Design in Gabor Setting

We consider a point-to-point communication link with a single transmitter, a wireless channel and a single receiver. We will first discuss TF foundations which is required to understand the waveform design. We consider the transmitted signal to be $x(t)$ and received signal to be $y(t)$.

In general, the transmitted signal $x(t)$ is affected by both multipath propagation and Doppler frequency shifts. The channel is linear time variant with channel operator $\mathscr{H}$. Received signal can be given as,

$$y(t) = \int_{-\infty}^{\infty}\int_{-\infty}^{\infty} \mathscr{S}_{\mathscr{H}}(\tau,\nu)(\mathbb{M}_\nu\mathbb{D}_\tau x)(t)d_\tau d_\nu, \tag{2.26}$$

where, $\mathscr{S}_{\mathscr{H}}(\tau,\nu)$ is delay-Doppler domain spreading function since it describes the spreading of transmit signal in time and frequency. AWGN channel can be obtained by $\mathscr{S}_{\mathscr{H}}(\tau, \nu) = \delta(\tau, \nu)$. Another description of LTV channel is time-varying impulse response, $h(t, \tau)$. The received signal can be given as,

$$y(t) = \int_{\infty}^{\infty} h(t, \tau)(\mathbb{D}_\tau x)(t) d_\tau. \tag{2.27}$$

2.5.1 Digital Communication in Gabor System

Let $d_{m,k}$ represents complex modulated data symbols where $m = 0, 1, \cdots, M-1$ represents time slots and $k = 0, 1, \cdots, N-1$ represents frequency slots. Let us suppose, we have total $W \times T$ resources, where W is the total bandwidth available in Hz and T is the total time available in seconds. Let T_s and F_s denote symbol duration and sub-carrier bandwidth. A lattice corresponds to an algebraic set which contains the coordinates of the filters in the TF plane [26]. In other words, it is a set generated by sampling the time-frequency plane. It determines the bandwidth efficiency and the reconstruction properties of a multi-carrier scheme. Without loss of generality, a lattice Λ can be described by a nonunique generator matrix, $\mathscr{L}$ as,

$$\mathscr{L} = \begin{bmatrix} \alpha & \beta \\ 0 & \theta \end{bmatrix}, \tag{2.28}$$

where $\alpha, \theta \neq 0$. The generator matrix contains the coordinates of the first two identifying points of the lattice in its column vectors, i.e., $(0, \alpha)$ and (β, θ). The locations of other points are calculated by applying $\mathscr{L}$ to $[m \; k]^{\mathrm{T}}$. For rectangular lattice, $\mathscr{L} = \begin{bmatrix} T & 0 \\ 0 & F \end{bmatrix}$, whereas for hexagonal lattice $\mathscr{L} = \begin{bmatrix} T & 0.5T \\ 0 & F \end{bmatrix}$. For a particular pair of (m, k), let, $[\mu_1^{(m,k)} \; \mu_2^{(m.k)}]^{\mathrm{T}} = \mathscr{L}[m \; k]^{\mathrm{T}}$, where, $\mu_1^{(m,k)}$ and $\mu_2^{(m.k)}$ are time and frequency shifts for a given value of (m, k).

Let $g(t)$ represents a prototype pulse shape. Transmitted signal $x(t)$ can be written as,

$$\begin{aligned} x(t) &= \sum_{m=0}^{M-1} \sum_{k=0}^{N-1} d_{m,k} g(t - \mu_1^{(m,k)}) e^{j2\pi \mu_2^{(m,k)}} \\ &= \sum_{m=0}^{M-1} \sum_{k=0}^{N-1} d_{m,k} (\mathbb{M}_{\mu_2^{(m,k)}} \mathbb{D}_{\mu_1^{(m,k)}} g(t)) \\ &= \sum_{m=0}^{M-1} \sum_{k=0}^{N-1} d_{m,k} g_{m,k}(t) \end{aligned} \tag{2.29}$$

Lattice density signifies the compactness of TF distribution of data symbols which is defined as $\varphi(\Lambda) = \frac{1}{|det(\mathscr{L})|}$. Spectral efficiency of a Gabor system is defined by, $\rho = \kappa \varphi(\Lambda)$, where κ is bits per symbol of the modulation scheme. Based on the lattice density the Gabor system can be categorized into three categories which are given as below.

- Under-sampled System ($\varphi(\Lambda) < 1$) : In this case any Gabor system is incomplete. The closed linear space spanned by $g_{m,k}(t)$'s is a proper subset of $L^2(\mathbb{R})$. However, well-localized prototype filters can be utilized.
- Critical-sampled System ($\varphi(\Lambda) = 1$) : In this case Gabor system is complete. The closed linear space spanned by $g_{m,k}(t)$'s is $L^2(\mathbb{R})$. Well localized prototype filters cannot be used according to Bailian–Low theorem [27].
- Over-sampled System ($\varphi(\Lambda) > 1$): In this case any Gabor system is overcomplete. $g_{m,k}(t)$'s constitute frames for $L^2(\mathbb{R})$. Well localized filters can be used.

When $x(t)$ is passed through a LTV channel, received signal, $y(t)$ at the transmitter can be given as,

$$
\begin{aligned}
y(t) &= (\mathscr{H}x)(t) + w(t) \\
&= \sum_{m=0}^{M-1}\sum_{k=0}^{N-1} d_{m,k}(\mathscr{H}g_{m,k})(t) + w(t)
\end{aligned}
\tag{2.30}
$$

Let, $\gamma_{m,k}(t)$ be the dual pulse of Gabor system at the receiver. These dual pulses can be obtained by TF shift of a prototype pulse shape i.e., $\gamma_{m,k}(t) = (\mathbb{M}_{\mu_2^{(m,k)}}\mathbb{D}_{\mu_1^{(m,k)}}\gamma)(t)$ [28]. Hence, the problem is simplified to find only prototype dual pulse. Now, estimated symbol $\hat{d}_{m',k'}$ can be found by correlating $y(t)$ to $\gamma_{m',k'}(t)$ i.e.,

$$
\begin{aligned}
\hat{d}_{m',k'} &= \langle \sum_{m=0}^{M-1}\sum_{k=0}^{N-1} d_{m,k}(\mathscr{H}g_{m,k})(t) + w(t), \gamma_{m',k'}(t)\rangle \\
&= \sum_{m=0}^{M-1}\sum_{k=0}^{N-1} d_{m,k}\langle(\mathscr{H}g_{m,k})(t), \gamma_{m',k'}(t)\rangle + \langle w(t), \gamma_{m',k'}(t)\rangle \\
\hat{d}_{m',k'} &= d_{m',k'}\langle(\mathscr{H}g_{m',k'})(t), \gamma_{m',k'}(t)\rangle \\
&\quad + \sum_{\substack{m=0 \\ m\neq m'}}^{M-1}\sum_{\substack{k=0 \\ k\neq k'}}^{N-1} d_{m,k}\langle(\mathscr{H}g_{m,k})(t), \gamma_{m',k'}(t)\rangle + \langle w(t), \gamma_{m',k'}(t)\rangle
\end{aligned}
\tag{2.31}
$$

Now, let us define, $n \in \mathbb{R}^2 = [m\ k]^{\mathrm{T}}\ \forall m, k$. $\zeta_{n,n_1} = \langle(\mathscr{H}g_{m,k})(t), \gamma_{m',k'}(t)\rangle$ and $\zeta_{n,n} = \langle(\mathscr{H}g_{m,k})(t), \gamma_{m,k}(t)\rangle$. Let I be a set which holds all vectors n corresponding to $\forall m, k$. Now, let $a = |\zeta_{n,n}|^2$ and $b = \sum_{n\in\{I\},n\neq n_1} |\zeta_{n,n_1}|^2$. SINR can be defined as,

$$
SINR = \frac{E_{\mathscr{H}}[a]}{\sigma_w^2 + E_{\mathscr{H}}[b]},
\tag{2.32}
$$

where, σ_w^2 is noise variance and we assume $E[d_n d_{n_1}] = \delta(n - n_1)$. Further, $E_{\mathscr{H}}[a]$ and $E_{\mathscr{H}}[b]$ can be given as,

$$
E_{\mathscr{H}}[a] = \int_\tau\int_\nu \mathscr{C}_{\mathscr{H}}(\tau, \nu)|\mathscr{A}_{g,\gamma}(\tau, \nu)|^2 d_\tau d_\nu
\tag{2.33}
$$

$$E_{\mathscr{H}}[b] = \sum_{n \in I, n \neq 0} \int_{\tau} \int_{\nu} \mathscr{C}_{\mathscr{H}}(\tau, \nu) |\langle g(t - \tau)e^{-j2\pi\nu t}, \gamma_n(t) \rangle|^2 d\tau d\nu, \qquad (2.34)$$

where, $\mathscr{A}_{x,y}(\tau, \nu) = \int_t x(t)y(t - \tau)e^{-j2\pi\nu t}$, is cross ambiguity function. It can be noted that both $E_{\mathscr{H}}[a]$ and $E_{\mathscr{H}}[b]$ are independent of m and k. Thus, on an average, all lattice points have the same SINR. The main objective of the waveform design over LTV channel is to choose transmitter pulse, receiver pulse and lattice such that SINR is maximized constrained on $\varphi(\Lambda)$ is a constant.

2.5.2 Waveform Design of Rectangular Lattice

Here, we consider a rectangular lattice i.e., $\mathscr{L} = \begin{bmatrix} T & 0 \\ 0 & F \end{bmatrix}$. For rectangular lattice, $g_{m,k}(t) := g(t - mT)e^{j2\pi kFt}$ and $\gamma_{m,k}(t) := \gamma(t - mT)e^{j2\pi kFt}$. We also consider that $\|g\|_{L^2(\mathbb{R})}^2 = 1$. Using these definitions, (2.3) and (2.31), $\hat{d}_{m', k'}$ can be obtained as,

$$\hat{d}_{m', k'} = \sum_{m=0}^{M-1} \sum_{k=0}^{N-1} q(m, k, m', k') d_{m,k}, \qquad (2.35)$$

where,

$$q(m, k, m', k') = e^{-j2\pi mT(k'-k)F}$$
$$\times \int_{\tau} \int_{\nu} \mathscr{S}_{\mathscr{H}}(\tau, \nu) \mathscr{A}_{g,\gamma}((m' - m)T - \tau, (k' - k)F - \nu)e^{j2\pi[\nu(mT+\tau)]-k'F\tau} d\tau d\nu$$
$$(2.36)$$

At this point, let us consider a channel with a bounded scattering function having the area of $\tau_{max} \times 2\nu_{max}$. It can be concluded that for estimation of $(m', k')^{\text{th}}$ data symbol, we require,

1. Maximum auto-term i.e.,

$$q(m', k', m', k') = \int_{\tau} \int_{\nu} \mathscr{S}_{\mathscr{H}}(\tau, \nu) \mathscr{A}_{g,\gamma}(-\tau, -\nu)e^{j2\pi[\nu(m'T+\tau)]-k'F\tau} d\tau d\nu = 1$$
$$(2.37)$$

2. Zero cross-term i.e.,

$$q(m, k, m', k')|m \neq m', k \neq k'$$
$$= e^{-j2\pi mT(k'-k)F} \int_{\tau} \int_{\nu} \mathscr{S}_{\mathscr{H}}(\tau, \nu) \mathscr{A}_{g,\gamma}((m' - m)T - \tau, (k' - k)F - \nu)$$
$$e^{j2\pi[\nu(mT+\tau)]-k'F\tau} d\tau d\nu = 0. \qquad (2.38)$$

$g(t)$ and dual $\gamma(t)$ are called ideal if they satisfy biorthogonality property for $(m', k')^{\text{th}}$ data symbol, $m' \in \mathbb{N}[0 \ M - 1]$ and $k' \in \mathbb{N}[0 \ N - 1]$ i.e.

$$\mathscr{A}_{g,\gamma}(t, f)|_{t=nT+(0 \ \tau_{max}), \ f=m\Delta f+(-\nu_{max} \ \nu_{max})} = \delta(n)\delta(m)U[0 \ \tau_{max}]U[-\nu_{max} \ \nu_{max}],$$
$$(2.39)$$

where, $U[a \ b](x) = \begin{cases} 1 \text{ If } a < x < b \\ 0 \text{ otherwise.} \end{cases}$. To satisfy biorthogonality property, cross ambiguty function should be confined in a rectangular area of $2\tau_{max}\nu_{max}$ which is impossible for a general wireless channel because of

1. considerations of under-spread wireless channel for which $2\tau_{max}\nu_{max} < 1$ [23],
2. and time-frequency uncertainty principle [29] which states that the minimum area covered by a pulse in TF domain is one which is satisfied by Gaussian pulse..

Thus, it can be concluded that ideal pulse does not exist. For example, if we consider a multi-carrier system with $\Delta f = 15$ KHz, carrier frequency, $f_c = 4$ GHz, $M = 128$ and $N = 512$. We take a 3GPP vehicular channel EVA [30] with vehicular speed of 500 Kmph. For this system $\tau_{max} = 2.5\mu s$, $\nu_{max} = 2.7$ KHz, and thus, $2\tau_{max}\nu_{max} = 0.0135 << 1$. It can be concluded that maintaining zero SIR is not possible for a general time-varying wireless channel. So, the researchers have tried to maximize SINR for transmission over time-varying channels.

SINR maximization, in general, is separated in two parts; (1) maximization of auto-term i.e., $E_{\mathscr{H}}[a]$, and (2) minimization of cross-term i.e., $E_{\mathscr{H}}[b]$. Following steps are recommended for maximization of SINR in [23].

- For maximization of $E_{\mathscr{H}}[a]$, $g(t)$ is chosen, whose ambiguity function is close to channels scattering function, i.e.,

$$g_{opt} = \arg \min_{g} ||\mathscr{C}_{\mathscr{H}} - |\mathscr{A}_g|^2||_{\mathbb{L}^2(\mathbb{R}^2)}^2. \tag{2.40}$$

Such problems are known as matching problems in radar literature.
- Next, γ is found by using biorthogonality condition i.e., $\mathscr{A}_{g,\gamma}(mT, kf) = \delta(m)\delta(k)$.
- For minimization of ISI and ICI, grid parameters are found by using matching rule,

$$\left(\frac{T}{F}\right)_{opt} \approx \frac{\tau_{max}}{2\nu_{max}}. \tag{2.41}$$

- For optimum AWGN performance, $g(t)$ pulses should be orthogonal. But, orthogonality of $g(t)$ cannot be maintained for optimum time-varying channels. Nonorthogonality of $g(t)$ enhances noise at the receiver. It can be shown that

$$E[|\langle w(t), \gamma_{m',k'}(t)\rangle|^2] \leq \frac{\sigma_w^2}{\sigma_d^2 TF A_g}, \tag{2.42}$$

where, A_g is the lower Riesz bound of the normalized $g_{m,k}$ frames. When $g_{m,k}$'s are orthonormal, $A_g = 1$. Otherwise, $A_g < 1$, and its value decreases with degree of nonorthogonality.

2.5.2.1 Ideal Eigenfunction of $\mathscr{H}$

Now suppose, $g_{m,k}(t)$ is an eigenfunction of $\mathscr{H}$ i.e., $(\mathscr{H}g_{m,k})(t) = \lambda_{k,m}g_{m,k})(t)$ for $\forall k, m$. Signal in (2.31) can be rewritten as,

$$
\hat{d}_{m',k'} = d_{m',k'}\lambda_{m',k'}\langle g_{m',k'}(t), \gamma_{m',k'}(t)\rangle
$$

$$
+ \sum_{\substack{m=0 \\ m\neq m'}}^{M-1}\sum_{\substack{k=0 \\ k\neq k'}}^{N-1} d_{m,k}\lambda_{m,k}\langle g_{m,k}(t), \gamma_{m',k'}(t)\rangle + \langle w(t), \gamma_{m',k'}(t)\rangle \qquad (2.43)
$$

Now, for perfect reconstruction, following condition should be met,

$$
\langle g_{m,k}(t), \gamma_{m',k'}(t)\rangle = \frac{1}{\lambda_{m',k'}}\delta(m - m')\delta(k - k'). \qquad (2.44)
$$

Above condition is called Waxler–Raz biorthogonal condition [31] for gabor systems. One subcase of this is when $\mathscr{H} = \mathscr{I}$ i.e., channel is AWGN. In this case, $\lambda_{m,k} = 1$, $\forall(m, k)$. When, $\gamma(t) = g(-t)$, Gabor system is called orthogonal as in this case, pulses in the transmitter become orthogonal. When Gabor system follows condition in (2.44) and $\gamma(t) \neq g(-t)$, it is called biorthogonal. When Gabor system does not follow (2.44), it is called nonorthogonal.

Using ideal eigenfunction has following problems;

1. Maintaining $(\mathscr{H}g_{m,k})(t) = \lambda_{k,m}g_{m,k}(t)$ is not always possible as $\mathscr{H}$ is random operator [22]. This means that it is not possible to maintain (2.43) for all possible realizations of $\mathscr{H}$.
2. Suppose, we compute eigenfunction of $\mathscr{H}$ for each transmission instance. Generally, these eigenfunctions will not have any structure and make transmission computationally complex. Additionally, other parameters such as PAPR and OoB radiation are also important parameters which cannot be guaranteed in such case. Chang in his classical paper [32] considered this case.
3. Getting exact information of $\mathscr{H}$ may not be always possible.

Due to above reasons, researchers have investigated approximated eigenfunction of $\mathscr{H}$ in Gabor setting for underspread WSSUS channel assuming the knowledge of channel scattering function $\mathscr{C}_{\mathscr{H}}(\tau, \nu)$. Works in this direction are summarized in next subsection.

2.5.3 Approximate Eigen Function for LTV Channel

Matz et. al. in [33] have shown that any pulse shape which is localized in TF domain is an eigenfunction. Assuming the knowledge of σ_τ and σ_ν, symbol and carrier width can be optimized as, $\frac{T}{F} = \frac{\sigma_\tau}{\nu}$. This criteria is known as matching criteria and first difined in [34]. Gabor in 1946 showed that Gaussian pulse the most concentrated pulse in the TF domain in the sense that the area formed by it's TF spread is minimum i.e., $B_g D_g = 1/2$, where B_g and D_g are pulse shape spread in time and frequency, respectively. The main problem with the Gaussian pulse is that it is non-orthogonal. Authors in [34] and [35] orthogonalize the Gaussian function to achieve orthogonal transmission. Another approach to find localized pulse shape was taken by Slepian in [36]. Slepian found that first order prolate spheroidal function's energy is maximally concentrated in a given symbol and carrier duration. Like Gaussian, first order PSWF is also nonorthogonal. To circumvent non-orthogonality, Vahlin et al., have modified PSWF pulses in [37].

Design of waveform assuming exact knowledge of $\mathscr{C}_{\mathscr{H}}(\tau, \nu)$ was first introduced in [22]. Authors derived pulse shape and its dual which maximize SINR for rectangular lattice. Optimum lattice to maximize SINR for IOTA prototype pulse was derived in [26]. Hexagonal lattice is found to be optimum. Both pulse shape and lattice are jointly optimized for maximization of SINR in [38].

When $TF < 1$, Gabor system is called under-sampled which also violates Nyquist's sampling theorem. Communication in such setup is called faster than Nyquist (FTN) communication and first introduced in 1978 by Mazo [39]. In Gabor setting, Han et. al. [40] investigated FTN communication assuming Gaussian prototype pulse shape and rectangular lattice. They have not used dual pulse at the receiver which may further increase the SINR.

2.6 OFDM

Orthogonal frequency division multiplexing (OFDM) is a multi-carrier modulation technique. OFDM is an advanced form of frequency division multiplexing (FDM) where the frequencies multiplexed are orthogonal to each other and their spectra overlap with the neighbouring carriers. systems. OFDM is built on the principle of overlapping orthogonal subcarriers. The frequency domain view of the signal is shown in Figure 2.4. The peak of one subcarrier coincides with the nulls of the other subcarriers due to the orthogonality. Thus, there is no interference from other subcarriers at the peak of a desired subcarrier even though the subcarrier spectrum overlap. It can be understood that OFDM systems avoid the loss in bandwidth efficiency prevalent in system using nonorthogonal carrier set. This brings in huge benefit in spectral efficiency for OFDM systems over earlier systems.

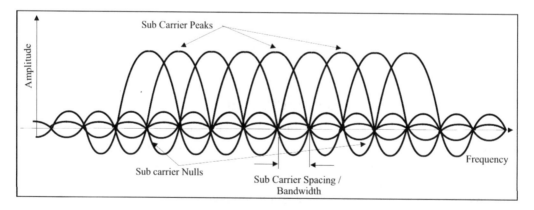

Figure 2.4   Orthogonal subcarriers in OFDM.

Table 2.1 Works on computing approximate eigenfunction of $\mathcal{H}$ in Gabor setting.

Work	Type of $\mathcal{C}_{\mathcal{H}}(\tau, \nu)$	transmit type	lattice	Problem Solved	Eigenfunction	Additional Notes
[29]	$\delta(\tau)\delta(\nu)$ AWGN	Nonorthogonal	Rectangular	Pulse shape which covers minimum area in TF plane., where $g(t) = min(B_g D_g)$., where B_g and D_g are time and frequency spread of $g(t)$ respectively.	Gaussian	
[32]	Complete Knowledge of $\mathcal{H}$	Orthogonal	Rectangular	Characteristics of optimum filter were derived for given $\mathcal{H}$	exact eigenfunction are derived at the transmitter	
[36]	$\delta(\tau)\delta(\nu)$ AWGN	Orthogonal	Rectangular	With a given time-frequency bound which pulse shape which covers minimum area in TF plane.	First order Prolate Spheroidal Wave Function (PSWF)	
[34]	Rectangular and Bounded	Orthogonal	Rectangular	Orthogonalization of Gaussian Pulse	IOTA Pulse	Matching Rule was induced. $\frac{\sigma_\tau}{\sigma_\nu} = \frac{T}{F}$, and $(TF)^{-1} = \varphi(\Lambda)$
[35]	Square and bounded	Orthogonal	Rectangular	Design of orthogonal pulse shapes which have maximum energy distribution within symbol duration and bandwidth	Pulse shape is linear combination of Hermite function	
[23]	ellipse and bounded	Biorthogonal	rectangular	Find prototype pulse shape and dual shape such that SINR is maximized for given scattering function	pulse shape and its dual are computed numerically	
[33]	unbounded but underspread	Non orthogonal	any	Computation of approximate Eigen function and values of $\mathcal{H}$	Any pulse shape which is well localized in TF domain	$E[\|y - (\mathcal{H}x)\|^2] \approx 4\pi^2 \sigma_\tau \sigma_\nu B_g D_g.$
[26]	Bounded	Orthogonal	Objective of this work	Finding optimum lattice	IOTA	Hexagonal Lattice is found to be optimum.
[38]	Bounded	Nonorthogonal	outcome of the work	Find prototype pulse shape dual shape and lattice such that SINR is maximized for given scattering function	Numerical evaluation	
[40]	Bounded	Nonorthogonal	Rectangular	Maximization of Spectral Efficiency	Gaussian Pulse is assumed	Faster than Nyquist
[41]	Bounded	Nonorthogonal	Hexagonal	Maximization of spectral efficiency	Linear combination of Hermite Functions	Spectral efficiency of 0.99 is achieved.

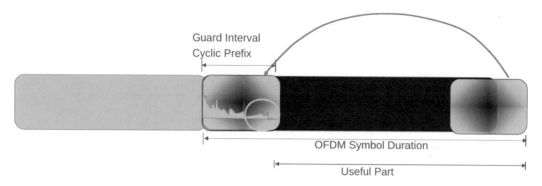

Figure 2.5   Illustration of the role of CP addition. It can be observed that ISI from adjacent symbol is absorbed in CP duration.

We begin by considering a block of QAM modulated symbols $\mathbf{d} = [d_0\ d_1 \cdots d_{N-1}]^{\mathrm{T}}$. We assume that data symbols are independent and identical i.e., $E[d_l d_l^*] = \sigma_d^2$, $\forall l$ and $E[d_l d_q^*] = 0$ when $l \neq q$. Let the total bandwidth B be divided into N number of subcarriers where symbol duration $T = \frac{N}{B}$ second and $\frac{B}{N}$ Hz is the subcarrier bandwidth. Transmitted OFDM signal, $x(n)$, $n \in [0, N-1]$, can be given as,

$$x(n) = \sum_{r=0}^{N-1} d_r e^{j2\pi \frac{rn}{N}}. \tag{2.45}$$

Alternatively, transmitted signal can be given as, $\mathbf{x} = \mathbf{W}_N \mathbf{d}$, where $\mathbf{W}_N$ is N-order inverse discrete Fourier transform (IDFT) matrix. Hence, transmitter can be implemented using N-point IDFT of complex data symbols, $\mathbf{d}$. IDFT can computed using low complexity radix-2 fast Fourier transform (FFT) algorithm. When OFDM signal is passed through multipath channel, symbol spreads in time. To avoid inter symbol interference (ISI), a cyclic prefix (CP) of length L_{CP} is added. Figure 2.5 illustrates the role of CP. Transmitted signal for OFDM-CP can be given as,

$$\mathbf{x}_{cp} = [x(N + L_{CP} - 1) : x(N - 1); \mathbf{x}]. \tag{2.46}$$

2.6.1 Channel

Let, $\mathbf{h} = [h_1, h_2, \cdots h_L]^{\mathrm{T}}$ be L length channel impulse response vector, where, h_r, for $1 \leq r \leq L$, represents the complex baseband channel coefficient of rth path [42], which we assume is zero mean circular symmetric complex Gaussian (ZMCSC). We also assume that channel coefficients related to different paths are uncorrelated. We consider, $N_{cp} \geq L$. Received vector of length $N_{CP} + NM + L - 1$ is given by,

$$\mathbf{y}_{cp} = \mathbf{h} * \mathbf{x}_{cp} + \boldsymbol{\nu}_{cp}, \tag{2.47}$$

where $\boldsymbol{\nu}_{cp}$ is AWGN vector of length $MN + N_{cp} + L - 1$ with elemental variance σ_ν^2.

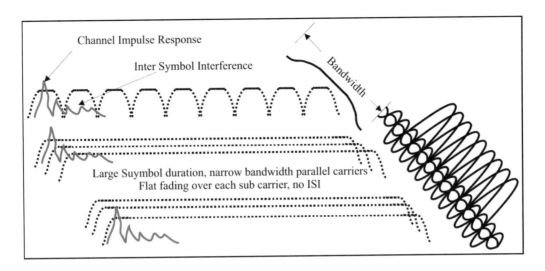

Figure 2.6 Time–frequency representation of OFDM signal.

2.6.2 Receiver

The first N_{cp} samples and last $L - 1$ samples of $\mathbf{y}_{cp}$ are removed at the receiver i.e., $\mathbf{y} = [\mathbf{y}_{cp}(N_{cp} + 1 : N_{cp} + MN)]$. Use of cyclic prefix converts linear channel convolution to circular channel convolution when $N_{cp} \geq L$[43]. The MN length received vector after removal of CP can be written as,

$$\mathbf{y} = \mathbf{H}\mathbf{W}_N\mathbf{d} + \boldsymbol{\nu}, \tag{2.48}$$

where $\mathbf{H}$ is circulant convolution matrix of size $MN \times MN$, which can be written as,

$$\mathbf{H} = \begin{bmatrix} h_1 & 0 & \cdots & 0 & h_L & \cdots & h_2 \\ h_2 & h_1 & \cdots & 0 & 0 & \cdots & h_3 \\ \vdots & & \ddots & & & & \\ h_L & h_{L-1} & \cdots & \cdots & \cdots & \cdots & 0 \\ 0 & h_L & \cdots & \cdots & \cdots & \cdots & 0 \\ \vdots & & \ddots & & & & \\ 0 & 0 & & h_L & \cdots & \cdots & h_1 \end{bmatrix}, \tag{2.49}$$

and $\boldsymbol{\nu}$ is WGN vector of length MN with elemental variance σ_ν^2.

As $\mathbf{H}$ is a circulant matrix, $\mathbf{H}$ can be factorized as, $\mathbf{H} = \mathbf{W}\boldsymbol{\Lambda}\mathbf{W}^{\mathrm{H}}$. $\boldsymbol{\Lambda} = \sqrt{N} \times diag\{\mathbf{W}_N^{\mathrm{H}}\mathbf{h}\}$, where, $\mathbf{h}$ is the first column of $\mathbf{H}$. This implies that $\boldsymbol{\Lambda}$ holds channel frequency samples which are sampled at the rate of $\frac{B}{N}$ samples per Hz in its diagonals.

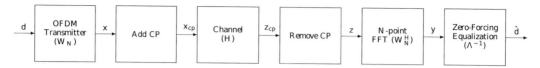

Figure 2.7   OFDM transceiver.

Hence, received vector after removal of CP can be written as,

$$\mathbf{y} = \mathbf{W}_N \mathbf{\Lambda} \mathbf{W}_N^H \mathbf{W}_N \mathbf{d} + \boldsymbol{\nu}$$
$$= \mathbf{W}_N \mathbf{\Lambda} \mathbf{d} + \boldsymbol{\nu} \tag{2.50}$$

Further, N-point FFT of $\mathbf{y}$ is computed i.e., $\mathbf{z} = \mathbf{W}_N^H \mathbf{y} = \mathbf{\Lambda d} + \mathbf{W}_N^H \boldsymbol{\nu}$. The time–frequency diagram of the OFDM signal in Figure 2.6 shows the difference between single and multi-carrier systems with respect to the symbol duration when compared against the channel impulse response. Single carrier system has a symbol duration which is decided by the sampling period of the system. When the channel impulse response is larger than this period, there is ISI. The whole bandwidth is split into a set of parallel orthogonal sub streams each of which has a long symbol duration. The symbol duration becomes significantly greater than the channel impulse response length. This makes each stream, i.e., each subcarrier, experiencing flat fading. Hence, a simple one tap equalizer per subcarrier can be used to equalize the effect of channel.

Effect of channel can be equalized using zero forcing (ZF) equalizer. Equalized OFDM vector $\hat{\mathbf{d}}$ can be written as,

$$\hat{\mathbf{d}} = \mathbf{\Lambda}^{-1} \mathbf{z} = \mathbf{d} + \mathbf{\Lambda}^{-1} \mathbf{W}_N^H \boldsymbol{\nu}. \tag{2.51}$$

OFDM transceiver can be understood in the light of Figure 2.7.

2.7 5G Numerology

OFDM with different numerology has been accepted as 5G waveforms by 3GPP [44]. 5G waveform is referred as new radio (NR). Numerology refers to subcarrier bandwidth, CP length and guard band. 5G supports two frequency ranges namely (1) FR1 : (450 MHz to 6000 MHz) in microwave frequency range and (2) FR2: (24250 MHz to 52600 MHz) in millimeter frequency range.

Definition of channel and transmission bandwidth is given in Figure 2.8. Different UE channel bandwidths as well as subcarrier bandwidth is supported within the same spectrum for transmitting to and receiving from UE's. Channel bandwidth of 5, 10, 15, 20, 25, 30, 40, 50, 60, 80, and 100 MHz are supported in FR1. Maximum transmission bandwidth allowed for FR1 and FR2 is given in Table 2.2 and 2.3 respectively.

Table 2.2 Maximum Transmission bandwidth Configuration for FR1. TBD means to be decided and NA means not applicable.

Channel Bandwidth (MHz) →	5	10	15	20	25	30	40	50	60	80	100
Subcarrier Bandwidth ↓ (KHz)	N_{RB}	N_{RB}	N_{RB}	N_{RB}	N_{RB}	N_{RB}	N_{RB}	N_{RB}	N_{RB}	N_{RB}	N_{RB}
15	25	52	79	106	133	TBD	216	270	NA	NA	NA
30	11	24	38	51	65	TBD	106	133	162	217	273
60	NA	11	18	24	31	TBD	51	65	79	107	135

Table 2.3 Maximum transmission bandwidth configuration for FR2. NA means not applicable.

Channel Bandwidth (MHz) →	50	100	200	400
Subcarrier Bandwidth ↓ (KHz)	N_{RB}	N_{RB}	N_{RB}	N_{RB}
60	66	132	264	NA
120	32	66	132	264

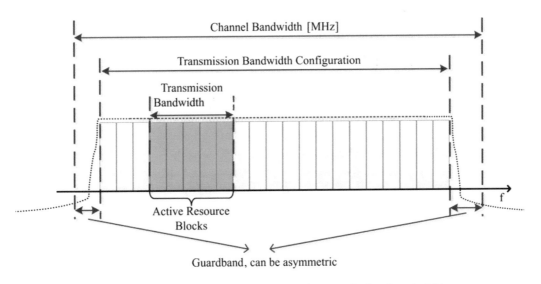

Figure 2.8   Definition of channel and transmission bandwidth.

2.7.1 Genesis

Use of different subcarrier bandwidth for OFDM was first proposed in 2005 [15] [13] [9]. Figure 2.9 illustrates the concept of variable subcarrier bandwidth in OFDM system. Variable subcarrier bandwidth OFDM implementation can be based on Band Division Multiplexing (BDM) as in Figure 2.10. The entire available bandwidth may be divided into subbands with different subcarrier bandwidth in each sub band for example a 100 MHz may be divided into chunks of 20 MHz, 10 MHz or 5 MHz. Each subband can be operated on by an IFFT with different number of subcarriers. The user equipment (UE) is assumed to require only one type of subcarriers bandwidth and hence will operate on only one subband. Therefore only one programmable FFT is needed for the user equipment. With a changing requirement of the subcarrier bandwidth the clock and the FFT size of the programable FFT may be dynamically configured. At the base station, as many FFTs may be used as there are different types of subcarrier bandwidths.

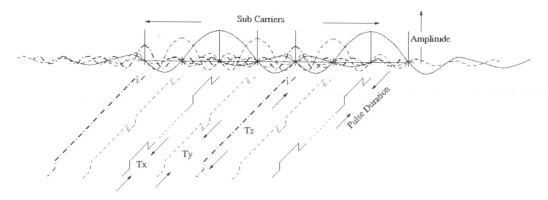

Figure 2.9　Time–frequency description of VSB-OFDM.

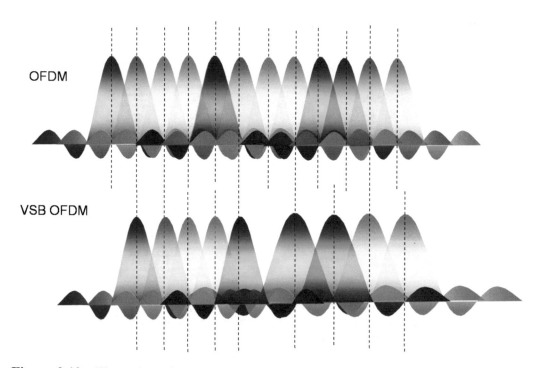

Figure 2.10　Illustration of VSB OFDM. As can be seen VSB OFDM accommodates different subcarrier bandwidth whereas subcarrier bandwidth, is fixed in OFDM.

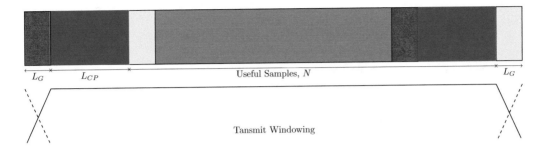

Figure 2.11   Windowed OFDM processing: Transmitter side.

2.8 Windowed OFDM

OFDM signal produces large side lobes in spectrum due to rectangular pulse shape in time domain signal. In Windowed CP-OFDM technique, time domain samples of CP-OFDM pulse are windowed to suppress the symbol energy at the edge of the CP-OFDM symbol[45]. Windowing transmitted OFDM symbols allow the amplitude to go smoothly to zero at the symbol boundaries leading to reduced discontinuity between symbols in time. This induces the spectrum of the transmitted signal to go down more rapidly. To maintain orthogonality among subcarriers additional guard samples are cyclically inserted. However, additional extended symbols decrease the spectrum efficiency.

2.8.1 Transmitter

We begin by considering OFDM-CP signal, $\mathbf{x}_{CP}$, given in (2.46). Windowed OFDM transmission can be understood We consider L_G to be the length of additional extended symbols. Last L_G samples of $\mathbf{x}$ are prefixed at the starting of the samples, $\mathbf{x}_{CP}$. Also, first L_G samples of $\mathbf{x}$ are postfixed at the end of the samples, $\mathbf{x}_{CP}$. Extended vector of length, $N + L_{CP} + 2L_W$, $\mathbf{x}_e$ can be given as, $\mathbf{x}_e = \mathbf{\Phi}\mathbf{x}$, where, $\mathbf{\Phi}$ can be given as,

$$
\mathbf{\Phi} = \begin{bmatrix} \mathbf{0}_{L_{CP}\times(N-L_{CP}-L_G)} & | & \mathbf{I}_{L_{CP}+L_G} \\ & \mathbf{I}_N & \\ \mathbf{I}_{L_G} | & \mathbf{0}_{L_{CP}\times(N-L_G)} & \end{bmatrix}_{(N+L_{CP}+2L_G)\times N} . \quad (2.52)
$$

Further, the extended symbols are smoothed using windowing. Let $\mathbf{g}$ be the L_G length row vector which holds the coefficients of pulse shape which is used to smooth first L_G samples of $\mathbf{x}_e$. Transmitted signal can be as, $\mathbf{x}_t = \mathbf{w}\circ\mathbf{x}_e$, where $\mathbf{w} = [\mathbf{g}\ \mathbf{1}_{N+L_{CP}}\ \mathbf{1}_{L_G}-\mathbf{g}]^{\mathrm{T}}$, $\mathbf{1}_-$ is a $(-)$ length row vector whose all entries are 1's, and $\circ$ represents Hadamard point-to-point vector multiplication. Adjacent symbols are overlapped in extended symbol duration to reduce the spectral efficiency loss. Spectral efficiency for windowed OFDM is $\frac{N}{N+L_{CP}+L_G}$ whereas for CP-OFDM spectral efficiency is $\frac{N}{N+L_{CP}}$. The transmitter processing can be understood in the light of Figure 2.11.

2.8.2 Receiver

At the receiver, symbols in extended period are discarded to eliminate the effect of windowing at the transmitter. Further, CP symbols can be discarded and the signal can be written in the similar way as in (2.48). Further, receiver processing can be similar as in Sec. 2.6.2.

In case of multi-user interference (MUI), windowing at the receiver helps in reducing the MUI [46, 47]. Figure 2.12 explains the receiver signal processing steps. In case of windowing at the receiver, CP is not discarded at the first, however, symbols in extended period are discarded and signal is $\mathbf{y}_{cp}$ as in (2.47). $\mathbf{y}_{cp}$ is first windowed using a window having smooth edges to reduce multi-user interference [46, 47]. We consider window of length $N + 2L_{rx}$, where, $2L_{rx}$ is the length of smooth pulse shape at each edge. We define, K_o, to be the starting sample index of windowing, we may call it window offset, $K_o \leq L_{cp} - 2L_{rx}$. Let $\mathbf{g}_r$ be a $2L_{rx}$ length row vector which holds pulse shape coefficients which are used to smooth first few samples of $\mathbf{y}_{cp}$. The $N + 2L_{rx}$ length windowed vector can be given as, $\mathbf{y}_w = \mathbf{w}_{rx} \circ \mathbf{y}_{cp}$, where $\mathbf{w}_{rx} = [\mathbf{0}_{K_o} \ \mathbf{g} \ \mathbf{1}_{N-2L_{rx}} \ \mathbf{1}_{2L_{rx}} - \mathbf{g} \ \mathbf{0}_{L_{CP}-K_o-2L_{rx}}]^{T}$, where $\mathbf{0}_-$ is a $(-)$ length row vector whose all entries are 0s. It is evident that first and last $2L_{rx}$ samples of $\mathbf{y}_w$ are distorted due to windowing. Interestingly, samples of $\mathbf{y}_w$ are cyclic in nature. We exploit this property to get useful N samples can recovered from $N + 2L_{rx}$ samples of $\mathbf{y}_w$ by using the principle of overlap and add. To do so, first $2L_{rx}$ samples are added to the last $2L_{rx}$ samples. We collect first $2L_{rx}$ samples in a row vector, $\mathbf{r} = [\mathbf{y}_w((N - 1 - L_{CP} + K_o + L_{rx} + 1) \mod N) : \mathbf{y}_w((N - 1 - L_{CP} + K_o + 2L_{rx}) \mod N)] \circ \mathbf{g}_{rx}$, and last $2L_{rx}$ samples in a row vector, $\mathbf{s} = [\mathbf{y}_w((N-1-L_{CP}+K_o+L_{rx}+1) \mod N) : \mathbf{y}_w((N - 1 - L_{CP} + K_o + 2L_{rx}) \mod N)] \circ (\mathbf{1} - \mathbf{g}_{rx})$. Overlap and added vector, $\mathbf{t} = (\mathbf{r} + \mathbf{s})^{T} = [\mathbf{y}_w((N - 1 - L_{CP} + K_o + L_{rx} + 1) \mod N) : \mathbf{y}_w((N-1-L_{CP}+K_o+2L_{rx}) \mod N)]$. Thus, effect of receiver windowing is eliminated in $\mathbf{s}$. Vector $\mathbf{s}$ replace first and last $2L_{rx}$ samples. Further, first and last L_{rx} samples are discarded to obtain N useful data symbols. Obtained signal can be written in the similar way as in (2.48). Further, receiver processing can be similar as in Sec. 2.6.2.

2.9 Filtered OFDM

In Filtered-OFDM (F-OFDM) [48, 49], available band is divided into multiple subbands. Each subband may use different OFDM parameters optimized for the application such as frequency spacing, cyclic prefix, transmit time interval(TTI). Each subband spectrum is filtered to avoid inter-subband interference. The main advantage is that different users (subbands) do not need to be time synchronized as a result global synchronization is relaxed. F-OFDM can achieve desirable frequency localization while enjoying the benefits of

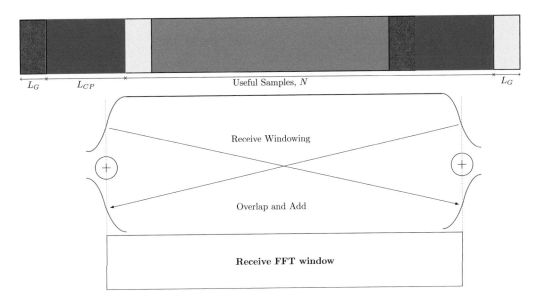

Figure 2.12 Windowed OFDM processing: Receiver side.

CP-OFDM and designing the filter appropriately. F-OFDM allows coexistence of different signal components with different OFDM primitives. By enabling multiple parameter configurations, F-OFDM is able to provide more optimum parameter numerology choice for each service group, and hence, better overall system efficiency. The impulse response of an ideal LPF is a sinc function which is infinitely long. Hence, for practical implementation, the sinc function is soft truncated with different window functions (1) Hann window (2) root-raised cosine window. In this way, the impulse response of the obtained filters will fade out quickly, and thus, limiting ISI introduced between consecutive OFDM symbols. Transmitter and receiver processing is demonstrated in Figure 2.13.

2.9.1 Transmitter

Let us consider a FOFDM system with N subcarriers that are divided into K subbands with each subband having M subcarriers, i.e. $N = MK$. Transmitted symbols can be given as,

$$\mathbf{d} = [\mathbf{d}_1 \ \mathbf{d}_2 \cdots \mathbf{d}_K]^{\mathrm{T}}, \tag{2.53}$$

where,

$$\mathbf{d}_k = [d_k(1) \ d_k(1) \cdots d_M]^{\mathrm{T}}, \tag{2.54}$$

is the transmitted signal for k^{th} sub-band, $1 \leq k \leq K$. Transmitted signal before filtering can be given as, $\mathbf{x}'_k = \mathbf{\Phi}' \mathbf{W}_k \mathbf{d}_k$, where $\mathbf{\Phi}'$ is CP addition matrix which can be obtained by putting $L_G = 0$ in (2.52) and $\mathbf{W}_k$ is $N \times M$ matrix which is obtained selecting $[(k-1)M+1]^{\mathrm{th}}$ to $(kM)^{\mathrm{th}}$ columns. Let $\mathbf{g}_k$ be the kth subband filter impulse response, which can be given as, $\mathbf{g}_k = [g_k(0) \ g_k(1) \cdots g_k(L_F - 1)]$, where L_F is the filter length. Let us

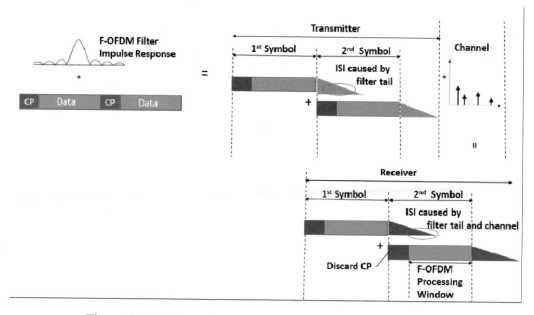

Figure 2.13 Transmitter and receiver processing of F-OFDM.

define a $(N + L_{CP} + L_F - 1) \times N + L_{CP}$ size teoplitz matrix, $\mathbf{G}_k$ whose first column is $[\mathbf{g}_k \ \mathbf{0}_{N+L_{CP}-1}]^{\mathrm{T}}$ and first row is $[g_k(0) \ \mathbf{0}_{N+L_{CP}-1}]^{\mathrm{T}}$. Transmitted signal for k^{th} subband can be given as,

$$\mathbf{x}_k = \mathbf{G}_k \mathbf{\Phi}' \mathbf{W}_k \mathbf{d}_k. \tag{2.55}$$

It is evident that transmitted signal is spread in time domain due to filtering as transmitted signal is $N + L_{CP} + L_F - 1$ long. As it is shown it Figure 2.13, the current symbol at k^{th} subband will overlap with next and previous symbol. Thus, ISI is introduced due to filter tail.

2.9.2 Receiver Processing

Let, $\mathbf{h}_k = [h_{(}0), \ h_k(1), \cdots h_k(L-1)]^{\mathrm{T}}$ be L length channel impulse response vector for k^{th} subband. Received vector of length $N + L_{CP} + L_F + L - 2$ is given by,

$$\mathbf{y}_k = \mathbf{h}_k * \mathbf{x}_k + \boldsymbol{\nu}_k, \tag{2.56}$$

where, $\boldsymbol{\nu}_k$ is AWGN vector of length $N + L_{CP} + L_F + L - 2$ with elemental variance σ_ν^2. Let us define a $(N + L_{CP} + L_F + L - 2) \times (N + L_{CP} + L_F - 1)$ teoplitz matrix, $\mathbf{H}'_k$ whose first row is $[h_k(0) \ \mathbf{0}_{N+L_{CP}+L_F-2}]$ and first column is $[\mathbf{h}_k \ \mathbf{0}_{N+L_{CP}+L_F-2}]^{\mathrm{T}}$. Let K_o be the offset in receiver windowing. $N + L_{CP}$ symbols are selected from $\mathbf{y}_k$ as $\mathbf{z}_k = \mathbf{y}_k[K_o \ : \ K_o + N + L_{CP} - 1]$. Let $\mathbf{H}_k$ be formed by selecting $(K_o + 1)^{\mathrm{th}}$ to $(K_o + N + L_C P)^{\mathrm{th}}$ row. Signal,$\mathbf{z}_k$ can be given as,

$$\mathbf{z}_k = \mathbf{H}_k \mathbf{x}_k + \tilde{\boldsymbol{\nu}}, \tag{2.57}$$

where, $\tilde{\nu}$ is $N + L_{cp}$ length noise vector. Next, $\mathbf{z}_k$ is passed through linear filtering, CP removal and DFT processing one by one. Processed signal can be given as,

$$\mathbf{z}_k^I = \mathbf{W}_k^H \mathscr{T} \mathbf{G}_k^H \mathbf{z}_k, \tag{2.58}$$

where, $\mathscr{T} = [\mathbf{0}_{N \times L_{CP}} \ \mathbf{I}_N]$ is CP removal matrix. $\mathbf{z}_k^I$ suffers from ISI due filter tail and channel dispersion. Let $\phi = \mathbf{W}_k^H \mathscr{T} \mathbf{G}_k^H \mathbf{G}_k \Phi' \mathbf{W}_k$. A zero forcing (ZF) receiver can be employed to combat ISI. ZF equalized signal can be given as,

$$\mathbf{z}_k^{zf} = (\phi^H \phi)^{-1} \phi^H \mathbf{z}_k^I. \tag{2.59}$$

2.10 Filter Bank Multi-Carrier

The idea of FBMC dates back to 1966, when Chang conceptualized cosine modulated tone (CMT) as a method to transmit PAM modulated data over band-limited multiple subcarriers at Nyquist rate [32]. A year later, in 1967, Stalzberg proposes a different version of FBMC which transmits QAM modulated data over band-limited multiple subcarriers at Nyquist rate [50], known as staggered modulated tone (SMT). The subcarrier distance is maintained at $\frac{1}{2T}$ (where T is the symbol duration), in CMT, to maintain Nyquist data rate and Vestigial side band modulation (VSB) is used to enforce orthogonality among subcarriers. Since, complex data symbol rides each subcarrier in SMT, the subcarrier distance of $\frac{1}{T}$ is used. This means each subcarrier is DSB-SC modulated. To maintain orthogonality, each complex symbol for a subcarrier is offset QAM modulated as well as adjacent subcarriers are staggered oppositely. However, recently, in 2010, Behrouz Farhang has shown that these two schemes are only slightly different and can be obtained via each other by a multiplication of a constant in frequency domain [51]. This also means that both CMT and SMT have similar performances in real transmission channels which is further emphasised in [51].

SMT can be implemented using polyphase structure of filter banks [52]. The polyphase approach has been the most used for FBMC as it provides a linear block approach. Using this structure, many receiver designs have been implemented, the most significant of which are by [53, 54]. However, the most efficient implementation of the FBMC based on frequency domain sampling was proposed by Bellanger in [55]. This implementation provides the simplest form of FBMC receiver in the frequency domain. Its performance has been analysed in [56, 57]. Also, the effect of timing and frequency offsets have been studied in [58] and equalizers for combating them have been designed in [59].

2.10.1 Cosine Modulated Tone

In this section, we will derive the continuous time domain CMT. Suppose, we have N parallel AM channels as depicted in Figure 2.14, which can transmit through a bandlimited transmission medium at a maximum possible data rate without ISI and ICI. The transmission filters $g_l(t)$ have gradual cut-off amplitude characteristics. The data rate per subcarrier is $\frac{1}{T}$, where T is the symbol duration. Hence, the subcarrier bandwidth is considered as $f_s = \frac{1}{2T}$. The center frequency for l^{th} subcarrier can be given as,

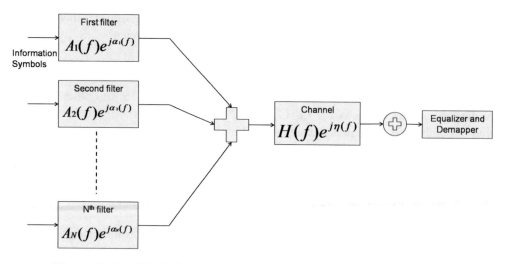

Figure 2.14 Block diagram of cosine modulated tone transmission.

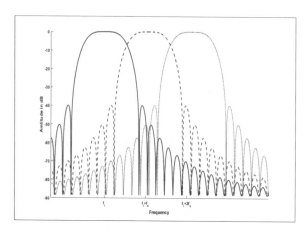

Figure 2.15 Overlapping Subcarriers in CMT.

$f_l = \frac{f_s}{2} + l f_s$. In the original work, Chang has considered a linear transmission medium but later works on CMT have consider an ideal channel. So to be in sync with existing literature, we are also considering an ideal channel. Let frequency response of lth transmission filter be $G_l(f)e^{j\alpha_l(f)}$, where $G_l(f)$ and $\alpha_l(f)$ is amplitude and phase response of lth transmission filter, where $l = 0, 1, \ldots, N-1$. Let $d_{m,l}$ be a real valued data symbol for $(m,l)^{th}$ time frequency slot, where $m = -\infty, \ldots, -1, 0, 1, \ldots, \infty$ and $l = 0, 1, \ldots, N-1$. Subcarriers are assumed to overlap into adjacent subcarriers as depicted in Figure 2.15.

Let us first concentrate on single subcarrier. For l^{th} subcarrier, received signal at the output of transmission medium can be given as,

$$y_l(t) = \sum_{m=-\infty}^{\infty} d_{m,l} g_l(t - mT) \qquad (2.60)$$

Now, the recieved signal overlap in time, they will be orthogonal if the signal corresponding to different data symbols are orthogonal i.e.,

$$\int_{-\infty}^{\infty} g_l(t)g_l(t-mT)dt = 0 \ \text{for} \ m = -\infty, \ldots, -1, 1, \ldots, \infty \ \text{and} \ k = 0, \ldots, N-1.$$

(2.61)

This is condition for ISI free communication. Now let us consider a different r^{th} subcarrier. The received signal can be given as,

$$y_r(t) = \sum_{m=-\infty}^{\infty} d_{m,r} g_r(t-mT)$$

(2.62)

Now, this signal from r^{th} subcarrier will overlap with signal from l^{th} subcarrier. These signals will be orthogonal if

$$\int_{-\infty}^{\infty} g_l(t)g_r(t-mT)dt = 0 \ \text{for} \ m = -\infty, \ldots, -1, 0, 1, \ldots, \infty$$

$$\text{and} \ l, r = 0, \ldots, N-1, \ l \neq r.$$

(2.63)

Using Fourier domain representation of $g_l(t)$, (2.61) can be rewritten as,

$$\int_{-\infty}^{\infty} G_l^2(f)e^{-j2\pi fmT}df = 0 \ \text{for} \ m = -\infty, \ldots, -1, 1, \ldots, \infty \ \text{and} \ l = 0, \ldots, N-1. \quad (2.64)$$

Let C be a real and positive constant and if $\int_{-\infty}^{\infty} C\cos(2\pi fmT)df = 0$, then this implies that $\int_{-\infty}^{\infty} C\sin(2\pi fmT)df = 0$. Since, $G_l(f)$ is real and positive, using this theorem, above condition for ISI free communication leads to

$$\int_{-\infty}^{\infty} G_l^2(f)\cos(2\pi fmT)df = 0 \ \text{for} \ m = -\infty, \ldots, -1, 1, \ldots, \infty \ \text{and} \ l = 0, \ldots, N-1.$$

(2.65)

In the above equation, since cosine is an even function around 0, values for positive k's will be same for negative k's. Apart from this $G_l(f)$ is an even function, the above equation can be simplified as,

$$\int_{0}^{\infty} G_l^2(f)\cos(2\pi fmT)df = 0 \ \text{for} \ m = 1, 2\ldots, \infty \ \text{and} \ l = 0, \ldots, N-1. \quad (2.66)$$

In the same way (2.63) can be written as,

$$\int_{-\infty}^{\infty} G_l(f)G_r(f)e^{-j(\alpha_l(f)-\alpha_r(f))2\pi fmT}df = 0 \ \text{for} \ m = -\infty, \ldots, -1, 0, 1, \ldots, \infty$$

$$\text{and} \ l, r = 0, \ldots, N-1, l \neq r. \quad (2.67)$$

Now, following the same steps as followed for the ISI free condition, ICI free condition can be given as,

$$\int_{0}^{\infty} G_l(f)G_r(f)\cos(\alpha_l(f)-\alpha_r(f))\cos(2\pi fmT)df = 0$$

(2.68)

$$\int_0^\infty G_l(f)G_r(f)\sin(\alpha_l(f)-\alpha_r(f))\sin(2\pi fmT)df = 0 \tag{2.69}$$

where, $m = 0, 1, \ldots, \infty$ and $l, r = 0, \ldots, N-1, l \neq r$.

2.10.2 Filter Characteristics

For ISI and ICI free communication, the transmitter filter $g_l(t)$, $\forall l$, should satisfy equations (2.66), (2.68) and (2.69). Let us assume that the transmitter filters are spread only to the adjacent subcarriers i.e.,

$$G_l(f) = \begin{cases} \Gamma_l(f) & f_l - f_s \leq f \leq f_l + f_s \\ 0 & \text{otherwise} \end{cases} \tag{2.70}$$

The ISI free condition given in (2.66), can be written as,

$$\int_{f_l-f_s}^{f_l+f_s} \Gamma^2(f)\cos(\frac{\pi fm}{f_s})df = 0 \text{ for } m = 1, 2\ldots, \infty \text{ and } l = 0, \ldots, N-1. \tag{2.71}$$

The cosine term in above equation, $\cos(\frac{\pi fm}{f_s})$ is even around $f_l - \frac{f_s}{2}$ as well as $f_l + \frac{f_s}{2}$. To make the equation zero, the function $\Gamma^2(f)$ should be odd around $f_l - \frac{f_s}{2}$ as well as $f_l + \frac{f_s}{2}$. But $\Gamma_l^2(f)$ is always greater than zero, hence, it can be written as, $\Gamma_l^2(f) = C_l + Q_l(f)$, where, C_l and $Q_l(f)$ are a positive constant and a pure AC component for l^{th} subcarriers respectively. Now, it can be very easily shown that, $\int_{f_l-f_s}^{f_l+f_s} C_l \cos(\frac{\pi fm}{f_s})df = 0$, $\forall m, l$. Hence, to satisfy (2.71), $Q_l(f)$ should be odd about $f_l - \frac{f_s}{2}$ and $f_l + \frac{f_s}{2}$, i.e.,

$$Q_l[(f_l + \frac{f_s}{2}) + f'] = -Q_l[(f_l + \frac{f_s}{2}) - f'], \ 0 < f' < \frac{f_s}{2} \tag{2.72}$$

$$Q_l[(f_l - \frac{f_s}{2}) + f'] = -Q_l[(f_l - \frac{f_s}{2}) - f'], \ 0 < f' < \frac{f_s}{2} \tag{2.73}$$

Next consider ICI, using the assumption $G_l(f) = 0$, for $f < f_l - f_s$, $f > f_l + f_f s$, it can be shown that

$$G_l(f)G_r(f) = 0 \text{ for } r = l \pm 2, l \pm 3, \ldots. \tag{2.74}$$

This implies that ICI conditions given in (2.68) and (2.69) are satisfied for $r = l \pm 2, l \pm 3, \ldots, \pm\infty$. It remains to investigate the ICI conditions for $r = l \pm 1$. Now, let us consider the case, $r = l + 1$. The equation (2.68) can be written as,

$$\int_0^\infty G_l(f)G_{l+1}(f)\cos(\alpha_l(f)-\alpha_{l+1}(f))\cos(2\pi fmT)df \tag{2.75}$$

$$= \int_{f_l}^{f_l+f_s} \sqrt{(C_l + Q_l(f))}\sqrt{(C_{l+1} + Q_{l+1}(f))}\cos(\alpha_l(f)-\alpha_{l+1}(f))\cos(\frac{\pi fm}{f_s})df.$$

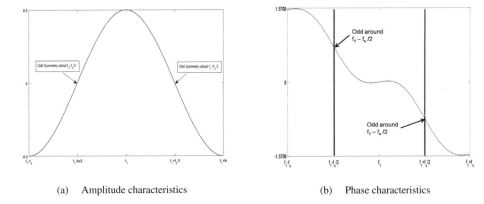

(a)   Amplitude characteristics   (b)   Phase characteristics

Figure 2.16   Filter characteristics.

in the above equation, the term, $\cos(\frac{\pi f m}{f_s})$ is even around $f_l + \frac{f_s}{2}$ and it can be very easily shown that $\sqrt{(C_l + Q_l(f))}\sqrt{(C_{l+1} + Q_{l+1}(f))}$ is an even function around $f_i + \frac{f_s}{2}$ (using (2.72)and 2.73). It is required that $\cos(\alpha_l(f) - \alpha_{l+1}(f))$ is odd around $f_i + \frac{f_s}{2}$ for zero ICI condition. Let, $\alpha_l(f) - \alpha_{l+1}(f) = \theta + \gamma_l(f)$, where $\theta_l \in \{0, 2\pi\}$ is a constant angle and $\gamma_l(f)$ is an arbitrary function. Let $f_t \in \{-\frac{f_s}{2}, \frac{f_s}{2}\}$, be a frequency variable, it is needed that $\cos(\theta + \gamma_l(f_l + \frac{f_s}{2} + f_t)) = -\cos(\theta + \gamma_l(f_l + \frac{f_s}{2} - f_t))$. Since, cosine function is odd around $\pm\frac{\pi}{2}$, take, $\theta = \pm\frac{\pi}{2}$. It can be easily verified that $\cos(\alpha_l(f) - \alpha_{l+1}(f))$ will be odd around $f_l + \frac{f_s}{2}$ iff $\gamma_l(f)$ is odd around $f_l + \frac{f_s}{2}$. In a similar manner, for $r = l - 1$, it can be written as $l = r + 1$, the same conditions can be obtained for r^{th} sub-carrier. In conclusion, for zero ICI condition, the phase characterstics, $\alpha_l(f)$ should be shaped such that

$$\alpha_l(f) - \alpha_{l+1}(f) = \pm\frac{\pi}{2} + \gamma_l(f), \quad f_l < f < f_l + f_s, \quad l = 1, 2, \ldots, N - 1, \quad (2.76)$$

where, $\gamma_l(f)$ is an arbitrary phase function with odd symmetry around $f_l + \frac{f_s}{2}$. Figure 2.16 illustrates the filter characteristics.

2.10.3 Simplified Filter Characteristics

In the previous section, filter characteristics, for communication over multiple band-limited subcarrier without ISI and ICI, are derived and can be followed using (2.72),(2.73) and (2.76). In this section, these conditions are simplified which are stated below.

1. C_l is same for all l.
2. $Q_l(f)$ are identical pulse shaped, i.e. ,

$$Q_{l+1}(f) = Q_l(f - f_s), \quad l = 0, 1, \ldots, N - 1. \quad (2.77)$$

Using this and above assumption it is straightforward to show that $G_l(f)$'s are identically pulse shaped, i.e.,

$$G_{l+1}(f) = G_l(f - f_s), \quad l = 0, 1, \ldots, N - 1. \tag{2.78}$$

The amplitude response of lth filter can be given as,

$$G_l(f) = G_0(f - lf_s), \tag{2.79}$$

where, $\alpha_0(f)$ is the amplitude response of the filter corresponding to first subcarrier ($N = 0$) centred at $f_0 = \frac{1}{4T}$.

3. $Q_l(f)$ is even around f_l, i.e.,

$$Q_l(f_l + f') = Q_l(f_l - f'), \quad 0 < f' < f_s, \quad l = 0, 1, \ldots, N - 1. \tag{2.80}$$

This also means that $G_l(f)$ is also even around f_l, i.e.

$$G_l(f_l + f') = G_l(f_l - f'), \quad 0 < f' < f_s, \quad l = 0, 1, \ldots, N - 1. \tag{2.81}$$

4. $\alpha_l(f)$ can be given as,

$$\alpha_l(f) = l\frac{\pi}{2} + \alpha_0(f - lf_s), \quad l = 0, 1, \ldots, N - 1, \tag{2.82}$$

where, $\alpha_0(f)$ is the phase response of the filter corresponding to first subcarrier ($N = 0$) centred at $f_0 = \frac{1}{4T}$. Since, $g_l(t)$ is real, $\alpha_l(f)$ is odd around f_l. Now, zero ICI condition given in (2.76) is satisfied if and only if $\alpha_l(f)$ is odd around $f_l + \frac{f_s}{2}$ and $f_l - \frac{f_s}{2}$. A special case of this will be when $\alpha_0(f) = 0$ and subsequently $\alpha_l(f) = l\frac{\pi}{2}$ for $l = 0, 1, \ldots, N - 1$. This condition is fulfilled when $g_0(t)$ satisfies symmetric condition i.e., $g_0(t) = g_0(-t)$.

Figure 2.17 shows power spectral density of different filters which fulfils above mentioned filter characteristics. Phydyas [60] filter has the lowest OoB radiation and generally preferred for FBMC transmission.

2.10.4 MMSE Equalizer for FBMC

Receiver design for FBMC is related to MMSE criterion [61, 62]. For this design, we can write the transmitted signal of FBMC in an alternate form as,

$$s[n] = \sum_{k=0}^{M-1} y_k[n] \tag{2.83}$$

where, the output for each subcarrier $y_k[n]$ is given as,

$$y_k[n] = \sum_{i=0}^{M-1} \sum_{l=-\inf}^{\inf} q_{ki}[l]v_i[n-l] + \eta_k[n] \tag{2.84}$$

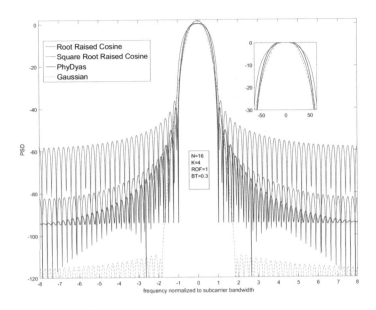

Figure 2.17 Prototype filters for CMT.

where, $v_i[n]$ denotes the OQAM input symbols and $q_{ki}[n]$ is given as,

$$q_{ki}[n] = (g_i \star h \star f_k)_{\downarrow \frac{M}{2}} \tag{2.85}$$

Here, $g_i[n]$ and $f_k[n]$ are the transmitter and receiver filters for the corresponding subcarrier indices. This equation represents represents the total signal processing involved in one subcarrier chain and the output obtained can be equalized directly. The equalizer mentioned here is a linear filter whose coefficients are designed according to the MMSE criterion. At first, L_{eq} samples of the output are collected and the filter coefficients are then multiplied with it to obtain the output. The output can be represented as,

$$y_k[n] = \sum_{i=k-1}^{k+1} Q_{ki}v_i[n] + B_k\eta[nM/2] \tag{2.86}$$

where Q_{ki} is given as,

$$Q_{ki} = \begin{bmatrix} q_{ki}[0] & q_{ki}[1] & \cdots & q_{ki}[L_q] & 0 & \cdots & 0 \\ 0 & q_{ki}[0] & q_{ki}[1] & \cdots & q_{ki}[L_q] & \ddots & \vdots \\ \vdots & \ddots & \ddots & \ddots & \ddots & \ddots & 0 \\ 0 & \cdots & 0 & q_{ki}[0] & q_{ki}[1] & \cdots & q_{ki}[L_q] \end{bmatrix}$$

is the convolution matrix formed by elements of q_{ki}, represented as $[q_{ki}[0], q_{ki}[1], ..., q_{ki}[L_q]]$ where $L_q + 1$ is the length of the convolution filter. B_k is the convolution and downsampling matrix. For a particular subcarrier k, only the contribution of the adjacent subcarriers $k - 1$ and $k + 1$ are taken into account. So, there are only 3 terms in the summation instead of M. $y_k[n]$ can be written as $y_k = [y_k[n], y_k[n - 1], ...y_k[n - L_{eq} + 1]]$ which consists of L_{eq} samples. The input samples contributing to the output are $v_k[n] = [v_k[n], v_k[n - 1], ..., v_k[n - L_q - L_{eq} + 1]]$. This indicates that each sample at the output has a contribution from L_q input samples.

Equation 2.86 can also be written considering $v_i[n]$ as purely real symbols. This way, the purely imaginary symbols are converted into real and the $j = \sqrt{-1}$ factor is multiplied with the corresponding column of Q_{ki}. This results in,

$$y_k[n] = \sum_{i=k-1}^{k+1} \overline{Q}_{ki} \overline{v}_i[n] + B_k \eta[nM/2] \qquad (2.87)$$

where, $\overline{Q}_{ki}$ is the modified Q_{ki} with alternate purely real and purely imaginary columns and $\overline{v}_i[n]$ is purely real. If the equalizer is denoted by w_k, then the output after equalization is,

$$Re(\tilde{v}_k[n]) = Re(w_k^H y_k[n]) \qquad for \ k + n \ even \qquad (2.88)$$

$$Im(\tilde{v}_k[n]) = Im(w_k^H y_k[n]) \qquad for \ k + n \ odd \qquad (2.89)$$

The equalizer is designed according to MMSE criterion which is given as,

$$\min_{w_k} E[|e_k|^2] = \min_{w_k} E[|\overline{w}_k^T \overline{y}_k[n] - v_k[n - \delta]|^2] \qquad (2.90)$$

where, $\overline{w}_k[n] = \begin{bmatrix} Re(w_k[n]) \\ Im(w_k[n]) \end{bmatrix}$ and $\overline{y}_k[n] = \begin{bmatrix} Re(y_k[n]) \\ Im(y_k[n]) \end{bmatrix}$. The MMSE equalizer can be defined as,

$$\overline{w}_k^T = 1_{k,\delta}^T \overline{\overline{Q}}_{kk}^T \left(\sum_{i=k-1}^{k+1} \overline{\overline{Q}}_{ki} \overline{\overline{Q}}_{ki}^T + \frac{1}{SNR} \overline{B}_k \overline{B}_k^T \right)^{-1} \qquad (2.91)$$

Here, δ is the equalizer delay and it depends on the transmitter and receiver filter lengths and the channel length. $\overline{\overline{Q}}_{ki}$ is given as, $\overline{\overline{Q}}_{ki} = \begin{bmatrix} Re(\overline{Q}_{ki}) \\ Im(\overline{Q}_{ki}) \end{bmatrix}$ and $\overline{B}_k = \begin{bmatrix} Re(B_k) & -Im(B_k) \\ Im(B_k) & Re(B_k) \end{bmatrix}$.

2.11 Universal Filtered Multi-Carrier

In UFMC, instead of pulse shaping every subcarrier individually, a group of subcarriers are pulse shaped together to form a resource block(RB) in the frequency domain. Each RB has a nearly flat frequency response within it and can be obtained by simple IDFT operation and filtering. There is some overlap between the RBs but it is negligible due to high sidelobe attenuation. So, UFMC is a near perfect orthogonal system when operated under normal conditions. Since, the pulse shape is wider than FBMC, the time domain response is shorter in length. Hence, UFMC spectral efficiency is higher. Also, the symbols do not overlap in time and block-wise decoding is possible.

2.11.1 Structure of UFMC Transceiver

The UFMC Transceiver is shown in figure 2.18.Let R_B be the number of RBs with each having N_i subcarriers as input with i varying from 1 to R_B. The total number of subcarriers is $N = \sum_{i=1}^{R_B} N_i$. The filtering can be done independently for each RB, and hence, the filter length depends on the width of the RB which in turn depends on the number of subcarriers. The filter lengths should be such that

$$N_1 + L_1 - 1 = N_2 + L_2 - 1 = N_3 + L_3 + 1........ = N_{R_B} + L_{R_B} - 1$$

for all the RBs. In the transmitter, at first an N point IDFT(N is the number of subcarriers) is taken for the subcarriers in each RB. The input has N_i samples and the output has N samples. The IDFT matrix is a partial matrix of the full IDFT one with the columns corresponding to the indices of the subcarriers in the $i - th$ RB. After IDFT, the samples are convolved with the corresponding RB filter of length L_i. Hence, the total number of samples at output is $N + L_i - 1$. Here, the filter length is same for all blocks, and, hence the output signal length is of length $N + L - 1$. The filter is designed in such a way that there is a rapid fall in time domain in the first $\frac{L}{2}$ and last $\frac{L}{2}$ samples. This means that these samples can be considered to have negligible values and this serves as a padding between successive symbols. In other words, the time domain behaviour of UFMC is almost similar to OFDM, where in place of OFDM's Cyclic Prefix, we have Zero Padding in UFMC. So, as long as the channel length is less than the filter length, there will be negligible or no ISI for both cases.

The filter used is an odd order Dolph–Chebyshey filter which is designed according to the number of subcarriers in the RB[63]. The filter takes three arguments, length, passband bandwidth and stopband attenuation which the can be designed according to the requirements and provides the filter impulse response. This gives a high level of control over the sdelobe attenuation for UFMC without having a large symbol length.

In the receiver, the received signal is first windowed using an appropriate filter(most preferably raised cosine filter) to contain it within a particular bandwidth.It is then zero-padded to make the total length $2N$. Then a $2N$ point DFT is taken. In the frequency domain, the $2N$ samples are alternatively selected, i.e., only the even indexed samples are selected. These samples carry the original data multiplied with the frequency domain channel coefficient. The odd-indexed samples carry the interference due to imperfect size of the DFT($2N$ point DFT taken over $N + L - 1$ length symbol). The channel is then equalized in the frequency domain.

2.11.2 System Model for UFMC

Here, the UFMC transceiver is expressed as a linear model. A simpler form of a linear model is given in [64] but here, we consider the detailed structure. The received signal after

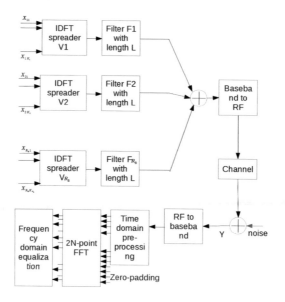

Figure 2.18   UFMC transceiver structure.

passing through the channel is given as,

$$\mathbf{Y} = \mathbf{H}_{(N+L+L_{ch}-2)\times(N+L-1)}\,\mathbf{\Phi}_{(N+L-1)\times((N+L-1)R_B)}$$
$$\mathbf{F}_{((N+L-1)R_B)\times NR_B}\,\mathbf{V}_{NR_B\times N}\mathbf{X}_{N\times 1} + \eta \qquad (2.92)$$

where $\mathbf{X}$ is the input data and is given as,

$$\mathbf{X} = [X_{11}, X_{12}, ..., X_{1K_1}, X_{21}, X_{22}, ..., X_{2K_2}, ...X_{R_B 1}, ..., X_{R_B K_{R_B}}]^T \qquad (2.93)$$

Here, $K_1, K_2, ...K_{R_B}$ are the number of subcarriers in the R_B RBs. In this implementation, all RBs have same number of subcarriers.

$\mathbf{V}$ is the combined IDFT matrix which is a block diagonal matrix given as, $diag(\mathbf{V_1}, \mathbf{V_2}...........\mathbf{V_{R_B}})$ where each $\mathbf{V_i}$ is an IDFT matrix of size $N \times K_i$ where only those columns of an $N \times N$ IDFT matrix is taken which correspond to the subcarrier indices of the RB under consideration. V_i is represented as,

$$\mathbf{V_i} = \begin{bmatrix} 1 & 1 & 1 & \cdots & 1 \\ e^{\frac{j2\pi i_1}{N}} & e^{\frac{j2\pi i_2}{N}} & e^{\frac{j2\pi i_3}{N}} & \cdots & e^{\frac{j2\pi i_{K_i}}{N}} \\ \vdots & \vdots & \vdots & \vdots & \vdots \\ e^{\frac{j2\pi(N-1)i_1}{N}} & e^{\frac{j2\pi(N-1)i_2}{N}} & e^{\frac{j2\pi(N-1)i_3}{N}} & \cdots & e^{\frac{j2\pi(N-1)i_{K_i}}{N}} \end{bmatrix} \qquad (2.94)$$

where $i_1, i_2............i_{K_i}$ corresponds to the indices in the particular $i-th$ RB.

F denotes the filtering matrix which is of $((N + L - 1)R_B) \times NR_B$. It is also a block diagonal matrix where each subblock is a toeplitz matrix denoting convolution. The filter length L is taken as same for all RBs here. The sublock $\mathbf{F_i}$ is given as,

$$
\mathbf{F_i} = \begin{bmatrix}
f_0 & 0 & 0 & 0 & \cdots & 0 & 0 \\
f_1 & f_0 & 0 & 0 & \cdots & 0 & 0 \\
\vdots & \vdots & \vdots & \vdots & \ddots & \vdots & \vdots \\
f_{L-1} & f_{L-2} & \cdots & f_0 & 0 & \cdots & 0 \\
\vdots & \vdots & \vdots & \vdots & \vdots & \vdots & \vdots \\
0 & 0 & \cdots & f_{L-1} & \cdots & f_0 & 0 \\
\vdots & \vdots & \vdots & \vdots & \ddots & \vdots & \vdots \\
0 & \cdots & \cdots & \cdots & \cdots & 0 & f_{L-1}
\end{bmatrix}
\tag{2.95}
$$

Here, $f_0, f_1, \ldots f_{L-1}$ are the filter coefficients. The filter coefficients are obtained by shifting the prototype filter to the center carriers of each RB. Thus, the coefficients are given as,

$$
[f_0 \; f_1 \; \cdots \; f_{L-1}] = f = [a_0 \; a_1 \; \ldots\ldots \; a_{L-1}] * exp(j \frac{(i-1) * PartFFTsize + (PartFFTsize/2)}{FFTsize})
\tag{2.96}
$$

Here, $[a_0, a_1, a_2 \ldots\ldots\ldots a_{L-1}]$ are the coefficients of the prototype filter.

$\mathbf{\Phi}$ is a concatenation of identity matrices which is given as,

$$
\mathbf{\Phi} = [\mathbf{I}_{(N+L-1)\times(N+L-1)} \quad \mathbf{I}_{(N+L-1)\times(N+L-1)} \quad \ldots\ldots\ldots\mathbf{I}_{(N+L-1)\times(N+L-1)}].
\tag{2.97}
$$

So, it consists of R_B identity matrices of size $N + L - 1$. Its purpose is to add the outputs of the RB together.

H is the channel matrix of size $(N + L + L_{ch} - 2) \times (N + L - 1)$ and it is also a Toeplitz matrix. It is given as,

$$
\mathbf{H} = \begin{bmatrix}
h_0 & 0 & 0 & 0 & \cdots & 0 & 0 \\
h_1 & h_0 & 0 & 0 & \cdots & 0 & 0 \\
\vdots & \vdots & \vdots & \vdots & \ddots & \vdots & \vdots \\
h_{L_{ch}-1} & h_{L_{ch}-2} & \cdots & h_0 & 0 & \cdots & 0 \\
\vdots & \vdots & \vdots & \vdots & \vdots & \vdots & \vdots \\
0 & 0 & \cdots & h_{L_{ch}-1} & \cdots & h_0 & 0 \\
\vdots & \vdots & \vdots & \vdots & \ddots & \vdots & \vdots \\
0 & \cdots & \cdots & \cdots & \cdots & 0 & h_{L_{ch}-1}
\end{bmatrix}
\tag{2.98}
$$

η is the additive white Gaussian noise vector.

Thus, the whole transmitter can be formulated as a linear model with the input as the data in the subcarriers and output in the time domain.

2.11.3 Output of the Receiver for the UFMC Transceiver Block Diagram

An analysis has been provided for the decoded symbols in the frequency domain at the receiver. By the decoding technique through $2N$-point FFT the effective ICI from the subcarriers within the corresponding RB and all the other RBs at the even subcarriers at the output of the $2N$-point FFT become zero.

The received signal without noise is given as,

$$\mathbf{Y_{1 \times (N+L-1)}} = \sum_{i=1}^{B} (\mathbf{f_i} * \mathbf{x_i}) \tag{2.99}$$

where, B is the number of RBs, f_i is the time domain filter for the $i-th$ RB. x_i is the IDFT of the data in the i-th RB and is given as,

$$x_i(l) = \frac{1}{N} \sum_{k \in S_i} X_i(k) exp(\frac{j2\pi lk}{N}) \tag{2.100}$$

where, S_i is the set of subcarrier indices belonging to the i-th RB. If we consider the output of only the $i-th$ RB then, the $m-th$ element of the output is,

$$y_i(m) = \sum_{g=0}^{L-1} f_i(g) x_i(m-g) \tag{2.101}$$

At the output, we take a $2N$-point DFT and thus, the $k'-th$ element is given as,

$$Y_i(k') = \sum_{m=0}^{2N-1} y_i(m) e^{\frac{-j2\pi mk'}{2N}} \tag{2.102}$$

$$Y_i(k') = \sum_{m=0}^{2N-1} \sum_{g=0}^{L-1} f_i(g) x_i(m-g) e^{\frac{-j2\pi mk'}{2N}} \tag{2.103}$$

$$Y_i(k') = \frac{1}{N} \sum_{m=0}^{2N-1} \sum_{k \in S_i} \sum_{g=0}^{L-1} f_i(g) X_i(k) e^{\frac{j2\pi(m-g)k}{N}} e^{\frac{-j2\pi mk'}{2N}} \tag{2.104}$$

$$Y_i(k') = \frac{1}{N} \sum_{m=0}^{2N-1} \sum_{k \in S_i} X_i(k) F_i(k) e^{\frac{j2\pi mk}{N}} e^{\frac{-j2\pi mk'}{2N}} \tag{2.105}$$

Now if $k' = 2p$ or k' is even for $p \in S_i$, then we have,

$$Y_i(2p) = \frac{1}{N} \sum_{m=0}^{2N-1} \sum_{k \in S_i} X_i(k) F_i(k) e^{\frac{j2\pi m(k-p)}{N}} \tag{2.106}$$

So, if $k = p$ then $Y_i(2p) = 2X_i(p)F_i(p)$ and for all other k, $Y_i(2p)$ is zero. Hence, the subcarrier data multiplied with the filter is recovered at twice the transmitter subcarrier index.

For $k' = 2p + 1$ or k' is odd for $p \in S_i$, then,

$$Y_i(2p + 1) = \frac{1}{N} \sum_{m=0}^{2N-1} \sum_{k \in S_i} X_i(k)F_i(k)e^{\frac{j2\pi mk}{N}} e^{\frac{-j2\pi m(2p+1)}{2N}} \tag{2.107}$$

In this equation, for any k, the data cannot be recovered in pure form and there is always interference from the other subcarriers in the block.

For the other RBs $j \neq i$, the particular output index for RB i will never match to any indices of the other RBs. Hence, output of equation 2.105 will always be zero for even subcarriers and interference terms for odd subcarriers.

2.12 Generalized Frequency Division Multiplexing (GFDM)

2.12.1 Introduction

In the previous sections, we presented a survey of a few legacy waveforms. We also presented methods to reduce the OoB in waveforms which can be divided into two categories (1) subcarrier pulse shaping wherein each sub-carrier is pulse shaped such as in FBMC, (2) a group of subcarrier pulse shaping wherein a group of subcarrier is pulse shaped such as in UFMC, and (3) transmitted signal shaping wherein signal before transmission is either windowed or filtered such as in W-OFDM and F-OFDM. We observed that subcarrier wise pulse shaping provides the lowest OoB radiation when compared with other methods. However, the former method can be further subdivided into two categories namely (i) linear pulse shaping (aka FBMC) wherein each subcarrier is linearly pulse shaped, and (ii) circularly pulse shaping wherein each subcarrier is circularly pulse shaped.

While linear pulse shapes reduce the OoB leakage significantly, they are spread in time and can reduce symbol efficiency especially in the case of small data frames or packets. On the other hand, circular pulse shapes are contained in time as well as in frequency and can reduce OoB without any symbol efficiency loss. This motivated investigation into circular pulse shaped waveform for low OoB leakage, known as generalized frequency division multiplexing (GFDM). GFDM is also more resilient to carrier frequency offset than OFDM, which is the key requirement for machine type communications (MTC) which requires asynchronous communications. Circular pulse shaping enables small frame sizes in GFDM which is a facilitator for low latency communications.

In this section, we introduce generalized frequency division multiplexing (GFDM).

2.12.1.1 Chapter Conents

The main contents of this section are as follows:

1. The system model of GFDM system under frequency selective fading channel (FSFC) is developed. Parallel between GFDM and discrete Gabor transform is drawn.

2. A comparative study of GFDM with other waveforms is made. This survey brings out advantages and disadvantages of the waveforms.

3. Two important properties of GFDM system are shows (i) product of modulation matrix with its hermitian is a block circulant matrix with circulant blocks (BCCB) and (ii) product of circulant convolution channel matrix with the modulation matrix when multiplied with its hermitian is a block circulant matrix. Closed form expression of post-processing SINR for LMMSE-GFDM receiver by exploiting above mentioned properties of GFDM system is developed. Assuming that the residual interference to be Gaussian distributed, we further computed semi-analytical BER. We also derive closed form expression for SINR of joint-MMSE receiver for GFDM. We use this SINR expression to evaluate BER in AWGN as well as in FSFC.

2.12.2 GFDM System in LTI Channel

GFDM is a multi-carrier modulation technique with some similarity to OFDM. We begin by considering a block of QAM modulated symbols $\mathbf{d} = [d_0 \ d_1 \cdots d_{MN-1}]^T$. We assume that data symbols are independent and identical i.e., $E[d_l d_l^*] = \sigma_d^2, \forall l$ and $E[d_l d_q^*] = 0$ when $l \neq q$. Let the total bandwidth B be divided into N number of subcarriers where symbol duration $T = \frac{N}{B}$ second and $\frac{B}{N}$ Hz is the subcarrier bandwidth. In case of OFDM this leads to orthogonal subcarriers. Let the symbol duration, T, be one time slot. GFDM is a block based transmission scheme and we consider a block to have M such time slots. Hence, in one block there are N subcarriers $\times M$ timeslots $= NM$ QAM symbols.

2.12.2.1 Transmitter

The $MN \times 1$ data vector $\tilde{\mathbf{d}} = [\tilde{d}_{0,0} \cdots \tilde{d}_{k,m} \cdots \tilde{d}_{N-1,M-1}]^T$, where $k = 0 \cdots N - 1$ denote subcarrier index and $m = 0 \cdots M - 1$ indicates time slot index. The data vector, $\tilde{\mathbf{d}}$, is modulated using GFDM Modulator. Data $\tilde{d}_{k,m}$ is first upsampled by N, which is represented as,

$$\tilde{d}_{k,m}^{up}(n) = \tilde{d}_{k,m}\delta(n - mN), \ n = 0, 1, \cdots, MN - 1. \tag{2.108}$$

Now, this upsampled data is pulse shaped. Impulse response of pulse shaping filter is represented by $g(n)$. Its length is MN. The upsampled data $\tilde{d}_{k,m(n)}^{up}$ is circularly convoluted with such pulse shaping filter $g(n)$ and can be written as,

$$x_{k,m}^f(n) = \sum_{r=0}^{MN-1} \tilde{d}_{k,m}g(n-r)_{MN}\delta(r - mN) \ \ = \tilde{d}_{k,m}g(n-mN)_{MN}. \tag{2.109}$$

Now, the filtered data is upconverted to k^{th} subcarrier frequency and is given as,

$$x_{k,m}(n) = \tilde{d}_{k,m}g(n-mN)_{MN}e^{\frac{j2\pi kn}{N}} \ \ = \tilde{d}_{k,m}a_{k,m}(n), \tag{2.110}$$

where, $a_{k,m}(n) = g(n-mN)_{MN}e^{\frac{j2\pi kn}{N}}$ for $n = 0, 1, \cdots MN - 1$, is called the kernel of GFDM for k^{th} subcarrier and m^{th} time slot. Now, output samples of all subcarrier frequencies and time slots are added and the GFDM signal can be given as,

$$x(n) = \sum_{m=0}^{M-1}\sum_{k=0}^{N-1} x_{k,m}(n) = \sum_{m=0}^{M-1}\sum_{k=0}^{N-1} \tilde{d}_{k,m}a_{k,m}(n). \qquad (2.111)$$

If we let, $l = mN + k$ for $m = 0, 1 \cdots M - 1$ and $k = 0, 1 \cdots N - 1$. Then, we may write (2.111) as,

$$x(n) = \sum_{l=0}^{MN-1} a_l^N(n)\tilde{d}_l^N, \qquad (2.112)$$

where, superscript N is identifier of the specific mapping between index l and time slot and subcarrier index tuple (k, m) which takes N as scaler multiplier in the mapping. If we collect all output samples of GFDM signal in a vector called $\mathbf{x} = [x(0)\ x(1) \cdots x(MN - 1)]^{\mathrm{T}}$ and all samples of $a_l^N(n)$ is a vector called $\mathbf{a}_l^N = [a_l^N(0)\ a_l^N(1) \cdots a_l^N(MN - 1)]$. Then using (2.112), $\mathbf{x}$ can be written as,

$$\mathbf{x} = \sum_{l=0}^{MN-1} \mathbf{a}_l^N \tilde{d}_l^N \equiv \sum_{l=0}^{MN-1} \tilde{d}_l^N \times l^{\text{th}} \text{ column vector } \mathbf{a}_l^N = \mathbf{A}\tilde{\mathbf{d}}^N, \qquad (2.113)$$

where, $\mathbf{A} = [\mathbf{a}_0^N\ \mathbf{a}_1^N \cdots \mathbf{a}_{MN-1}^N]$ is modulation matrix , $\tilde{\mathbf{d}}^N = [\tilde{d}_0\ \tilde{d}_1 \cdots \tilde{d}_l \cdots \tilde{d}_{NM-1}]^{\mathrm{T}}$ is precoded data vector. If we collect all samples of $a_{k,m}(n)$ is vector called $\mathbf{a}_{k,m} = [a_{k,m}(0)\ a_{k,m}(0) \cdots \cdots a_{k,m}(MN - 1)]^{\mathrm{T}}$, then $\mathbf{A}$ can also be written as,

$$\mathbf{A} = [\underbrace{\mathbf{a}_{0,0}\ \mathbf{a}_{1,0} \cdots \mathbf{a}_{N-1,0}}_{1^{\text{st}}\text{timeslot}} | \underbrace{\mathbf{a}_{0,1}\ \mathbf{a}_{1,1} \cdots \mathbf{a}_{N-1,1}}_{2^{\text{nd}}\text{timeslot}} | \cdots | \underbrace{\mathbf{a}_{0,M-1}\ \mathbf{a}_{1,M-1} \cdots \mathbf{a}_{N-1,M-1}}_{M^{\text{th}}\text{timeslot}}]$$

$$= \begin{bmatrix} g(0) & g(0) & \cdots & \cdots & g(0) & g(MN-N) & \cdots & \cdots & g(MN-N) & \cdots \\ g(1) & g(1)\rho & \cdots & \cdots & g(1)\rho^{N-1} & g(MN-N+1) & \cdots & \cdots & g(MN-N+1)\rho^{N-1} & \vdots \\ \vdots & \vdots & \cdots & \cdots & \vdots & \vdots & \cdots & \cdots & \cdots & (M-2)N\text{terms} \\ \vdots & \vdots & \cdots & \cdots & \vdots & \vdots & \cdots & \cdots & \cdots & \vdots \\ g(MN-1) & g(MN-1)\rho^{MN-1} & \cdots & \cdots & g(MN-1)\rho^{(N-1)(MN-1)} & g(MN-N-1) & \cdots & \cdots & g(MN-N-1)\rho^{(N-1)(MN-1)} & \cdots \end{bmatrix},$$
$$\underbrace{}_{\mathbf{a}_{0,0}} \underbrace{}_{\mathbf{a}_{1,0}} \qquad \underbrace{}_{\mathbf{a}_{N-1,0}} \underbrace{}_{\mathbf{a}_{0,1}} \qquad \underbrace{}_{\mathbf{a}_{N-1,1}}$$
$$(2.114)$$

where, $\rho = e^{\frac{j2\pi}{N}}$. At this point, it is also interesting to look into the structure $\mathbf{a}_{k,m}$ vectors which constitute columns of $\mathbf{A}_N$. The first column of $\mathbf{A}_N$ i.e., $\mathbf{a}_{0,0}$ holds all coefficients of pulse shaping filter and other columns of $\mathbf{A}_N$ or other vectors in the set of $\mathbf{a}_{k,m}$'s are time and frequency shifted version of $\mathbf{a}_{0,0}$ where frequency index k denotes $\frac{k}{N}$ shift in frequency and time index m denotes m time slot or mN sample cyclic shift. Taking clue from description of column vectors $\mathbf{a}_{k,m}$, structure of $\mathbf{A}_N$ can be understood by (2.114). Columns having same time shift are put together. Columns which are N column index apart are time shifted version of each other. Columns having same time shift are arranged in increasing order of frequency shift. We will explore time domain and frequency domain

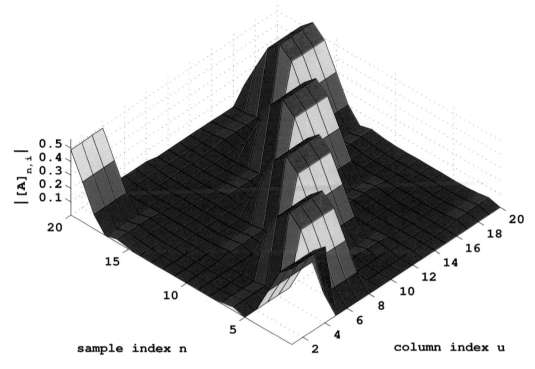

Figure 2.19 Time domain view of columns of $\mathbf{A}$ for $N = 4$, $M = 5$ and $ROF = 0.9$.

behaviour for an example $\mathbf{A}_N$ with total subcarriers $N = 4$ and total time slots $M = 5$. Pulse shaping filter is taken to be root raised cosine (RRC) with roll of factor (ROF) of 0.9. Figure 2.19 shows the absolute values of each column index $u = 0, 1 \cdots MN - 1$ with sample index $n = 0, 1 \cdots , MN - 1$. It can be observed that first $N = 4$ columns have zero time shift and next N columns have unit time shift and so on. Further, Figure 2.20 provides TF view of $\mathbf{A}$.

Alternatively, if we take $l = kM + m$ in equation (2.111), modulation matrix $\mathbf{A}_\mathrm{M}$ can be represented as,

$$\mathbf{A}_\mathrm{M} = [\underbrace{\overbrace{\mathbf{a}_{0,0}}^{1^{st}\text{ time}} \overbrace{\mathbf{a}_{0,1}}^{2^{nd}\text{ time}} \cdots \overbrace{\mathbf{a}_{0,M-1}}^{\mathtt{M}^{th}\text{ time}}}_{1^{st}\text{frequency}} | \underbrace{\mathbf{a}_{1,0}\ \mathbf{a}_{1,1} \cdots \mathbf{a}_{1,M-1}}_{2^{nd}\text{frequency}} | \cdots | \underbrace{\mathbf{a}_{N-1,0}\ \mathbf{a}_{N-1,1} \cdots \mathbf{a}_{N-1,M-1}}_{N^{th}\text{frequency}}]. \quad (2.115)$$

Matrix $\mathbf{A}_N$ and $\mathbf{A}_M$ can be related as $\mathbf{A}_N = \boldsymbol{\zeta}\mathbf{A}_M$, where $\boldsymbol{\zeta}$ is a permutation matrix which permutes column of matrix applied. Cyclic prefix is added to GFDM modulated block to prevent inter-block interference in FSFC. CP of length N_{CP} is perpended to $\mathbf{x}$. After adding CP, transmitted vector, $\mathbf{x}_{cp}$, can be given as

$$\mathbf{x}_{cp} = [\mathbf{x}(MN - N_{cp} + 1 : MN) \ ; \ \mathbf{x}] \quad (2.116)$$

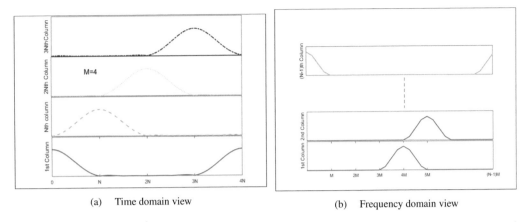

(a) Time domain view   (b) Frequency domain view

Figure 2.20 Time–frequency view of GFDM matrix **A**. $M = 5$.

In the rest of chapter, for equations which are valid for both $\mathbf{A}_M$ and $\mathbf{A}_N$, modulation matrix will be denoted as $\mathbf{A}$.

2.12.2.2 Self-interference in GFDM

As discussed earlier, $\mathbf{A}$ is a nonunitary matrix which means that columns of $\mathbf{A}$ matrix are not orthonormal to each other. Each data symbol is multiplied to a column of $\mathbf{A}$. Hence, data symbols will interfere to each other.

2.12.2.3 Receiver

Let, $\mathbf{h} = [h_1, h_2, \cdots h_L]^{\mathrm{T}}$ be L length channel impulse response vector, where, h_r, for $1 \leq r \leq L$, represents the complex baseband channel coefficient of rth path [42], which we assume is zero mean circular symmetric complex Gaussian (ZMCSC). We also assume that channel coefficients related to different paths are uncorrelated. We consider, $N_{cp} \geq L$. Received vector of length $N_{CP} + NM + L - 1$ is given by,

$$\mathbf{y}_{cp} = \mathbf{h} * \mathbf{x}_{cp} + \boldsymbol{\nu}_{cp}, \qquad (2.117)$$

where, $\boldsymbol{\nu}_{cp}$ is AWGN vector of length $MN + N_{cp} + L - 1$ with elemental variance σ_ν^2. The first N_{cp} samples and last $L - 1$ samples of $\mathbf{y}_{cp}$ are removed at the receiver i.e., $\mathbf{y} = [\mathbf{y}_{cp}(N_{cp} + 1 : N_{cp} + MN)]$. Use of cyclic prefix converts linear channel convolution to circular channel convolution when $N_{cp} \geq L$ [43]. The MN length received vector after removal of CP can be written as,

$$\mathbf{y} = \mathbf{HA}\tilde{\mathbf{d}} + \boldsymbol{\nu}, \qquad (2.118)$$

where, $\mathbf{H}$ is circulant convolution matrix of size $MN \times MN$, which can be written as,

$$\mathbf{H} = \begin{bmatrix} h_1 & 0 & \cdots & 0 & h_L & \cdots & h_2 \\ h_2 & h_1 & \cdots & 0 & 0 & \cdots & h_3 \\ \vdots & & \ddots & & & & \\ h_L & h_{L-1} & \cdots & \cdots & \cdots & \cdots & 0 \\ 0 & h_L & \cdots & \cdots & \cdots & \cdots & 0 \\ \vdots & & \ddots & & & & \\ 0 & 0 & & h_L & \cdots & \cdots & h_1 \end{bmatrix}, \tag{2.119}$$

and $\boldsymbol{\nu}$ is WGN vector of length MN with elemental variance σ_ν^2. Received vector $\mathbf{y}$ is distorted due to (i) Self interference as subcarriers are nonorthogonal [65][3] and (ii) Inter-carrier interference(ICI) due to FSFC.

Equalizers in GFDM can be categorized into two major categories namely (1) Linear equalizers and (2) Non-Linear equalizers. It is worthwhile to mention that the nonlinear equalizers can have better performance than linear equalizers at the cost of higher complexity. In general, nonlinear equalizers are iterative and have linear equalizers in its core. In this chapter, we focus on linear equalizers. For the discussion on non-linear receivers, readers can refer [68] [69] [70] [71].

Linear GFDM equalizers in multi-path channel can be broadly categorized as (a) two-stage equalizers[65, 72, 73] and (b) one-stage equalizers [74, 75, 69, 76]. In a two-stage receiver, channel equalization is followed by GFDM demodulation, while in the one stage receiver, the effect of channel and GFDM modulation is jointly equalized [74]. In next subsections, we will describe these equalizers.

2.12.2.4 Two Stage Equalizer

Channel equalized vector, $\mathbf{y}$, can be given as [77],

$$\mathbf{y} = \mathbf{W}_{MN}\boldsymbol{\Lambda}_{eq}\mathbf{W}_{MN}^{\mathrm{H}}\mathbf{z} \quad = a\mathbf{A}\mathbf{d} + \boldsymbol{b} + \tilde{\boldsymbol{\nu}}, \tag{2.120}$$

where, $\boldsymbol{\Lambda}_{eq} = \begin{cases} \boldsymbol{\Lambda}^{-1} & \text{for ZF FDE} \\ [\boldsymbol{\Lambda}^{\mathrm{H}}\boldsymbol{\Lambda} + \frac{\sigma_\nu^2}{\sigma_d^2}\mathbf{I}_{MN}]^{-1}\boldsymbol{\Lambda}^{\mathrm{H}} & \text{for MMSE FDE} \end{cases}$ where, $\tilde{\boldsymbol{\nu}} = \mathbf{W}_{MN}\boldsymbol{\Lambda}_{eq}\mathbf{W}_{MN}^{\mathrm{H}}\boldsymbol{\nu}$,

$$a = \begin{cases} 1 & \text{for ZF FDE} \\ \frac{1}{MN}\sum_{r=0}^{MN-1} \frac{|\tilde{h}(r)|^2}{|\tilde{h}(r)|^2 + \frac{\sigma_\nu^2}{\sigma_d^2}} & \text{for MMSE FDE,} \end{cases}$$

---

[3]In other words, matrix $\mathbf{A}$ is nonunitary. Authors in [66] proposed to use unitary $\mathbf{A}$ to eliminate self-interference. However, in this part we consider a nonunitary $\mathbf{A}$ to retain the flexibility of GFDM. Other way to remove self-interference is by using offset QAM modulated symbols [67]. This study only considers QAM modulated GFDM system.

b is residual interference, given in (2.121) and $\tilde{\boldsymbol{\nu}} = \mathbf{W}_{MN}\mathbf{\Lambda}_{eq}\mathbf{W}_{MN}^{\mathrm{H}}\boldsymbol{\nu}$ is post-processing noise.

$$b = \begin{cases} \mathbf{0} \quad \text{for ZF FDE} \\ [\mathbf{W}_{MN}\mathbf{\Lambda}_{eq}\mathbf{\Lambda}\mathbf{W}_{MN}^{\mathrm{H}} - \\ \frac{1}{MN}\sum_{r=0}^{MN-1}\frac{|\tilde{h}(r)|^2}{|\tilde{h}(r)|^2+\frac{\sigma_\nu^2}{\sigma_d^2}}\mathbf{I}_{MN}]\mathbf{A}\mathbf{d} \quad \text{for MMSE FDE} \end{cases} \tag{2.121}$$

Channel equalized vector, $\mathbf{y}$, is further equalized to remove the effect of self-interference. Estimated data, $\hat{\mathbf{d}}$, can be given as,

$$\hat{\mathbf{d}} = \mathbf{A}_{eq}\mathbf{y}, \tag{2.122}$$

where, $\mathbf{A}_{eq}$ is GFDM equalization matrix which can be given as,

$$\mathbf{A}_{eq} = \begin{cases} \mathbf{A}^{\mathrm{H}} \text{ for MF Equalizer} & \text{(2.123a)} \\ \mathbf{A}^{-1} \text{ for ZF Equalizer} & \text{(2.123b)} \\ [\mathbf{R}_{\boldsymbol{\nu}} + \mathbf{A}^{\mathrm{H}}\mathbf{A}]^{-1}\mathbf{A}^{\mathrm{H}} & \text{(2.123c)} \\ \text{for biased MMSE Equalizer,} \\ \mathbf{\Theta}gfdm^{-1}[\mathbf{R}_{\boldsymbol{\nu}} + \mathbf{A}^{\mathrm{H}}\mathbf{A}]^{-1}\mathbf{A}^{\mathrm{H}} & \text{(2.123d)} \\ \text{for unbiased MMSE Equalizer,} \end{cases}$$

where, $\mathbf{R}_{\boldsymbol{\nu}} = E[\tilde{\boldsymbol{\nu}}\tilde{\boldsymbol{\nu}}^{\mathrm{H}}]$ is noise correlation matrix after channel equalization. In the case of AWGN, $\mathbf{R}_{\boldsymbol{\nu}} = \frac{\sigma_\nu^2}{\sigma_d^2}\mathbf{I}$. For multipath fading channel, $\mathbf{R}_{\boldsymbol{\nu}}$ is a full matrix since the noise after channel equalization is colored [72]. $\mathbf{\Theta}gfdm^{-1}$ is a diagonal bias correction matrix for GFDM-MMSE equalizer, where,

$$\mathbf{\Theta}gfdm = diag\{[\frac{\sigma_\nu^2}{\sigma_d^2}\mathbf{I} + \mathbf{A}^{\mathrm{H}}\mathbf{A}]^{-1}\mathbf{A}^{\mathrm{H}}\mathbf{A}\}. \tag{2.124}$$

2.12.2.5 One-Stage Equalizer

One stage ZF equalizer is same as two-stage ZF equalizer. Hence, we consider only MMSE equalizer [4]. Since one-stage MMSE equalizes channel and self-interference jointly, we call it Joint-MMSE equalizer. Joint-MMSE equalizer can be of two types, namely, (1) biased-joint-MMSE and (2) unbiased-joint-MMSE. Equalized data symbol vector, $\hat{\mathbf{d}}_{\text{JP}}$, can be

---

[4]Authers in [76] proposed a Gabor tranform based one stage equalizer. But, its BER performance is shown to be worse than MMSE equalizer

given as, $\hat{\mathbf{d}}_{\text{JP}} = \mathbf{B}_{eq}\mathbf{y}$, where, $\mathbf{B}_{eq}$ is joint-MMSE equalizer matrix and can be given as,

$$
\mathbf{B}_{eq} = \begin{cases} [\frac{\sigma_{\nu}^2}{\sigma_d^2}\mathbf{I} + (\mathbf{HA})^{\text{H}}\mathbf{HA}]^{-1}(\mathbf{HA})^{\text{H}} \\ \text{for biased joint-MMSE Equalizer} \\ \mathbf{\Theta}^{-1}[\frac{\sigma_{\nu}^2}{\sigma_d^2}\mathbf{I} + (\mathbf{HA})^{\text{H}}\mathbf{HA}]^{-1}(\mathbf{HA})^{\text{H}} \\ \text{for unbiased joint-MMSE Equalizer,} \end{cases}
\tag{2.125}
$$

where, $\mathbf{\Theta}^{-1}$ is diagonal bias correction matrix for joint-processing, where, $\mathbf{\Theta} = diag\{[\frac{\sigma_{\nu}^2}{\sigma_d^2}\mathbf{I} + (\mathbf{HA})^{\text{H}}\mathbf{HA}]^{-1}(\mathbf{HA})^{\text{H}}\mathbf{HA}\}$.

It should be noted that in some cases, few data symbols $d_{k,m}$ can be pre-known, such as guard subcarriers or zero subsymbols to reduce adjacent channel interference [74]. Like authors in [78, 72, 75, 76], we treat all these pre-known data symbols as random variables. However, receiver performance can be improved by exploiting the knowledge of zero data symbols but at the cost of increased complexity. To throw some more light on this case, consider S_{za} to be the set which holds the zero time-frequency slots (m_z, k_z), where $m_z \in [0, M-1]$ and $k_z \in [0, N-1]$. Covariance matrix of transmit data $\mathbf{d}$ is, $\mathbf{C}_{dd} = \mathbf{I}_{MN} - \mathbf{1}_{S_{za}}$, where, $\mathbf{1}_{S_{za}}(p,q) = \begin{cases} 1 \text{ when } p = q = m_z k_z \\ 0 \text{ otherwise} \end{cases}$. Joint-MMSE matrix in such a cases is,

$$
\mathbf{B}_{eq} = \mathbf{C}_{dd}(\mathbf{HA})^{\text{H}}[\mathbf{HA}\mathbf{C}_{dd}(\mathbf{HA})^{\text{H}} + \frac{\sigma_{\nu}^2}{\sigma_d^2}\mathbf{I}_{MN}]^{-1}.
\tag{2.126}
$$

Now, let us consider a GFDM system with $N = 128$, $M = 5$ and ROF $=0.3$. We switch off 32 subcarriers on each edge i.e., total 320 data symbols are zero. When zero data symbols are treated as random variable, joint-MMSE receiver is implemented using $\mathbf{C}_{dd} = \mathbf{I}_{MN}$ which is equivalent to biased receiver in (2.125). BER performance of this system over urban micro (UMi) [30] channel is given in Figure 2.21. It is observed that (2.126) can have improved performance over (2.125).

2.12.3 GFDM in Gabor System

2.12.3.1 Discrete Gabor Transform

Consider a discrete signal x of length L is a finite sequence $[x(0)\ x(1)\ \cdots x(L-1)] \in \mathbb{C}^L$. The inner product of f and g is $\langle f, g \rangle_{l^2(\mathbb{N})} = \sum_{n=0}^{L-1} f(n)\bar{g}(n)$ and the norm is given by $\|f\| = \langle f, f \rangle_{l^2(\mathbb{N})}$. The sequence $x(n)$ is extended to infinite sequences of period L by $x(n + rL) = x(n)$ for $r \in \mathbb{Z}$. We define two operators on $l^2(\mathbb{N})$ namely (1) time-shift operator $\mathbb{D}_a x(n) = x(n - a)$ and (2) modulation operator $\mathbb{M}_b x(n) = x(n)e^{j2\pi bn/L}$, where $a, b \in \mathbb{Z}$. For a given Gabor atom γ the set of Gabor analysis functions $\{\gamma_{m,k}\}$ is given by $\gamma_{m,k} = \mathbb{M}_{kb}\mathbb{D}_{ma}\gamma$, for $m \in \mathbb{N}[0\ M-1]$, $k \in \mathbb{N}[0\ N-1]$, where $Nb = Ma = L$. The parameters a and b represents time and frequency sampling intervals, and are also referred as time and frequency shift parameters.

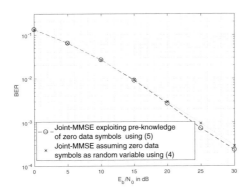

Figure 2.21 Uncoded BER vs $\frac{E_b}{N_0}$ for MMSE receivers using (2.125) and (2.126) in case of zero data symbols over Urban micro (UMi) channel [30]. Total 64 sub-carriers are switched off. N=128, M=5. 16 QAM modulation and raised cosine (RC) pulse with ROF = 0.3 is considered.

When $L = ab$, Gabor system is critically sampled whereas $L > ab$ and $L < ab$ represents oversampled and under-sampled Gabor system, respectively. Discrete Gabor transform of the function f can be given as,

$$d_{m,k} = \langle x, \gamma_{m,k} \rangle = \sum_{n=0}^{L-1} f(n)\bar{\gamma}_{m,\,k}(n)$$

$$= \sum_{n=0}^{L-1} x(n)\bar{\gamma}((n-ma) \mod L)e^{-j2\pi kb/L}, \; \forall m, \, k. \tag{2.127}$$

A more compact representation of analysis can be formed in matrix-vector multiplication notation. Consider, $\mathbf{d}$ be a MN length vector having $d_{m,k}$ as its $(k+Nm)^{\text{th}}$ element, whereas $\mathbf{x}$ is an L length vector having $x(r)$ as its rth element. Similarly, consider $\mathfrak{G}$ be a $L \times MN$ matrix whose $(k+Nm)^{\text{th}}$ column is $\gamma_{m,k}$. Gabor analysis can be given as,

$$\mathbf{d}_{\text{MN}\times 1} = \mathfrak{G}_{\text{L}\times \text{MN}}^{\text{H}} \mathbf{x}_{\text{L}\times 1}. \tag{2.128}$$

We can reconstruct x from its Gabor coefficients $d_{m,k}$ using a dual Gabor functions $g_{m,k} = \mathbb{M}_{kb}\mathbb{D}_{ma}g$ via

$$x(n) = \sum_{m=0}^{M-1}\sum_{k=0}^{N-1} d_{m,k}g_{m,k}. \tag{2.129}$$

Consider $\mathbb{Z}$ be a $L \times MN$ matrix whose $(k+Nm)^{\text{th}}$ column is $g_{m,k}$. The duality condition can be expressed as,

$$\mathfrak{G}\mathbb{Z}^{\text{H}} = \mathbf{I}_L. \tag{2.130}$$

Thus, $\mathbb{Z}$ can be computed by computing the inverse of $\mathfrak{G}$. In general, $\mathfrak{G}$ is a rectangular matrix, thus $\mathfrak{G}^{-1}$ does not exist. When $ab < L$ there are many dual functions satisfying condition (2.130). We can can compute the dual with minimum norm by computing pseudo-inverse of $\mathfrak{G}^\dagger = \mathfrak{G}^H[\mathfrak{G}\mathfrak{G}^H]^{-1}$. The frame operator $\mathbf{S}$ can be defined as,

$$\mathbf{S} = \mathfrak{G}\mathfrak{G}^H. \qquad (2.131)$$

The minimum norm dual satisfies $\mathbf{S}g = \gamma$. Thus, minimum norm dual can be computed using the inversion of a positive definite matrix $\mathbf{S}$.

2.12.3.2 Critically Sampled Gabor Transform

When, $b = M$ and $a = N$, $L = MN$; and Gobor system becomes critically sampled. The Gabor synthesis equation in (2.129) becomes

$$x(n) = \sum_{m=0}^{M-1} \sum_{k=0}^{N-1} d_{m,k}\, g[(n - mN) \mod MN]\, e^{j2\pi nk/N}. \qquad (2.132)$$

Comparing (2.132) with (2.111), it can be established that GFDM modulation is critically sampled inverse discrete Gabor transform. This means that $\mathbf{A}$ which is described in (2.114) can also be interpreted as a IDGT matrix. This opens up an opportunity to bring in the know-how of DGT for better understanding and easy implementation of GFDM systems.

2.12.4 Bit Error Rate Computation for MMSE Receiver

2.12.4.1 MMSE Receiver

To equalize the channel and GFDM induced self-interference, a joint MMSE equalizer [74] is considered here. Equalized data can be given as,

$$\hat{\mathbf{d}} = [\mathbf{I}_{MN}\frac{\sigma_{\nu 2}}{\sigma_{d2}} + (\mathbf{HA})^H(\mathbf{HA})]^{-1}(\mathbf{HA})^H\mathbf{y}$$
$$= \mathscr{B}\mathbf{d} + \mathscr{C}\boldsymbol{\nu}, \qquad (2.133)$$

where, $\mathscr{B} = [\mathbf{I}_{MN}\frac{\sigma_{\nu 2}}{\sigma_{d2}} + (\mathbf{HA})^H(\mathbf{HA})]^{-1}(\mathbf{HA})^H(\mathbf{HA})$ and $\mathscr{C} = [\mathbf{I}_{MN}\frac{\sigma_{\nu 2}}{\sigma_{d2}} + (\mathbf{HA})^H(\mathbf{HA})]^{-1}(\mathbf{HA})^H$. First term in the above equation holds desired plus interference values and second term holds the post processing noise values.

2.12.4.2 SINR Computation

Suppose, we want to detect l^{th} symbol. Estimated l^{th} symbol can be given as,

$$\hat{d}_l = [\mathscr{B}]_{l,l}d_l + \sum_{r=0,r\neq l}^{MN-1} [\mathscr{B}]_{l,r}d_r + \sum_{r=0}^{MN-1} [\mathscr{C}]_{l,r}\nu_r, \qquad (2.134)$$

where, first term is desired term, second term is interference term and third term is post processed noise term. Using above equation, $E[|\hat{d}_l|^2] = \sigma_d^2[\mathscr{B}]_{l,l}^2 + \sigma_d^2\sum_{r=0,r\neq l}^{MN-1} |[\mathscr{B}]_{l,r}|^2 +$

$\sigma_\nu^2 \sum_{r=0}^{MN-1} |[\mathscr{C}]_{l,r}|^2$, where first term is average signal power $P_{Sig,l}$, second term is average interference power $P_{Sig+Inr,l}$ and third term is average post processing noise power $P_{Npp,l}$. Using (2.133), $E[\hat{\mathbf{d}}\hat{\mathbf{d}}^H] = \sigma_d^2 \mathscr{B}\mathscr{B}^H + \sigma_\nu^2 \mathscr{C}\mathscr{C}^H$, where diagonal values of first matrix term holds average signal plus interference power $P_{Sig+Inr,l}$, diagonal values of second matrix term holds average post processing noise power, $\mathscr{B}\mathscr{B}^H = [[\mathbf{I}_{MN}\frac{\sigma_\nu 2}{\sigma_d 2} + (\mathbf{HA})^H(\mathbf{HA})]^{-1}(\mathbf{HA})^H(\mathbf{HA})]^2 = \mathscr{B}^2$ and $\mathscr{C}\mathscr{C}^H = [\mathbf{I}_{MN}\frac{\sigma_\nu 2}{\sigma_d 2} + (\mathbf{HA})^H(\mathbf{HA})]^{-1}(\mathbf{HA})^H(\mathbf{HA})[\mathbf{I}_{MN}\frac{\sigma_\nu 2}{\sigma_d 2} + (\mathbf{HA})^H(\mathbf{HA})]^{-1}$. Using this, average signal power $P_{Sig,l}$ and average interference power $P_{Inr,l}$, average signal plus interference power $P_{Sig+Inr,l}$ and post processed noise power $P_{Npp,l}$ for l^{th} symbol, can be given as,

$$P_{Sig,l} = \sigma_{d^2}|[\mathscr{B}]_{l,l}|^2, \ P_{Sig+Inr,l} = \sigma_{d^2}|[\mathscr{B}\mathscr{B}^H]_{l,l}|, \ P_{Inr,l}$$
$$= P_{Sig+Inr,l} - P_{Sig,l} \text{ and } P_{Npp,l} = \sigma_{\nu^2}|[\mathscr{C}\mathscr{C}^H]_{l,l}|. \quad (2.135)$$

SINR for l^{th} symbol can be computed as,

$$\gamma_l = \frac{P_{Sig,l}}{P_{Inr,l} + P_{Npp,l}} \quad (2.136)$$

Now, we will compute SINR for FSFC and AWGN channel separately.

2.12.4.3 Frequency Selective Fading Channel (FSFC)

Both $\mathscr{B}$ and $\mathscr{C}\mathscr{C}^H$ involve $(\mathbf{HA})^H(\mathbf{HA})$. To compute (2.136) we explore the product. $(\mathbf{HA})^H\mathbf{HA}$ is a Hermitian matrix and hence, it can be diagonalized as, $(\mathbf{HA})^H\mathbf{HA} = \mathbf{V}\Lambda\mathbf{V}^H$, where, $\mathbf{V}$ is a unitary matrix which holds eigenvectors of $(\mathbf{HA})^H\mathbf{HA}$ in it's columns and $\Lambda = diag\{\lambda_0, \lambda_1, \cdots \lambda_{MN-1}\}$ is a diagonal matrix which holds eigenvalues of $(\mathbf{HA})^H\mathbf{HA}$. We can write, $\mathscr{B} = \mathbf{V}\tilde{\Lambda}\mathbf{V}^H$, where, $\tilde{\Lambda} = diag\{\tilde{\lambda}_0, \tilde{\lambda}_1, \cdots \tilde{\lambda}_{MN-1}\}$, $\tilde{\lambda}_s = \frac{\lambda_s}{\frac{\sigma_\nu^2}{\sigma_d^2}+\lambda_s}$ and $\mathscr{C}\mathscr{C}^H = \mathbf{V}\tilde{\tilde{\Lambda}}\mathbf{V}^H$, where, $\tilde{\tilde{\Lambda}} = diag\{\tilde{\tilde{\lambda}}_0, \tilde{\tilde{\lambda}}_1, \cdots \tilde{\tilde{\lambda}}_{MN-1}\}$ and $\tilde{\tilde{\lambda}}_s = \frac{\lambda_s}{(\frac{\sigma_\nu^2}{\sigma_d^2}+\lambda_s)^2}$. Putting values of $\mathscr{B}$ and $\mathscr{C}$ into (2.135), we can get,

$$P_{Sig,l} = \sigma_{d^2}|[\mathbf{V}\tilde{\Lambda}\mathbf{V}^H]_{l,l}|^2, \ P_{Inr,l} = \sigma_{d^2}|\sum_{r=0,r\neq l}^{MN-1}[\mathbf{V}\tilde{\Lambda}\mathbf{V}^H]_{l,r}|^2 \text{ and } P_{Npp,l}$$
$$= \sigma_{\nu^2}|[\mathbf{V}\tilde{\tilde{\Lambda}}\mathbf{V}^H]_{l,l}|^2. \quad (2.137)$$

Through the reduction, complex expression involving matrix inverse as in (2.134), is brought to a simpler form (using eigenvalue decomposition of $(\mathbf{HA})^H\mathbf{HA}$) i.e., instead of computing $\mathscr{B}$ we can proceed directly with $\mathbf{HA}$.

2.12.4.4 Additive White Gaussian Noise Channel (AWGN)

In case of AWGN, $\mathbf{H} = \mathbf{I}_{MN}$, hence, we can write the following,

$$\mathscr{B} = [\frac{\sigma_\nu^2}{\sigma_d^2}\mathbf{I} + \mathbf{A}^H\mathbf{A}]^{-1}\mathbf{A}^H\mathbf{A}, \ \mathscr{C} = [\frac{\sigma_\nu^2}{\sigma_d^2}\mathbf{I} + \mathbf{A}^H\mathbf{A}]^{-1}\mathbf{A}^H \text{ and } \mathscr{C}\mathscr{C}^H$$

$$= [\frac{\sigma_\nu^2}{\sigma_d^2}\mathbf{I} + \mathbf{A}^H\mathbf{A}]^{-1}\mathbf{A}^H\mathbf{A}[\frac{\sigma_\nu^2}{\sigma_d^2}\mathbf{I} + \mathbf{A}^H\mathbf{A}]^{-1}. \tag{2.138}$$

It can be seen that $\mathbf{A}^H\mathbf{A}$ is a major component of the analysis. It is important to study the $\mathbf{A}^H\mathbf{A}$ before we proceed further. We will see the properties of $\mathbf{A}_N^H\mathbf{A}_N$ below.

$\mathbf{A}_N^H\mathbf{A}_N$

Matrix, $\mathbf{A}_N$, can be decomposed as,

$$\mathbf{A}_N = [[\mathscr{G}_0\mathbf{E}]_{MN \times N} \quad \cdots \quad [\mathscr{G}_{M-1}\mathbf{E}]_{MN \times N}]_{MN \times MN}, \tag{2.139}$$

where, $\mathscr{G}_p$'s be a set of $MN \times MN$ matrices, where $p = 0, \cdots, M-1$. Suppose the first matrix in the set $\mathscr{G}_0 = diag\{\mathbf{g}^T\}$, where $\mathbf{g} = [g(0)\ g(1)\cdots g(MN-1)]^T$. Any matrix in the set can be written as a circularly shifted version of $\mathscr{G}_0$ along its diagonals i.e., p^{th} matrix in the set can be written as $\mathscr{G}_p = \mathrm{diag}\{\mathrm{circshift}[\mathbf{g}^T, -pN]\}$, where circshift represents right circular shift operation. $\mathscr{G}_p$ can be described as,

$$\mathscr{G}_p = \begin{bmatrix} g(-pN)_{MN} & & \\ & \ddots & \\ & & g(-pN+MN-1)_{MN} \end{bmatrix}_{MN \times MN} \tag{2.140}$$

and, $\mathbf{E} = [\mathbb{W}_N \cdots M\ times \cdots \mathbb{W}_N]^T$ is a $MN \times N$ matrix, where $\mathbb{W}_N$ is $N \times N$ normalized inverse DFT matrix [79]. $[\mathbf{A}_N^H\mathbf{A}_N]_{r,q} = \mathbf{E}^H\mathscr{G}_r^H\mathscr{G}_q\mathbf{E}$, where $r, q = 0, \cdots, M-1$ and $\mathbf{E}^H\mathscr{G}_r^H\mathscr{G}_q\mathbf{E}$ is $N \times N$ matrix. $[\mathbf{A}_N^H\mathbf{A}_N]_{r,q}$ can be written as,

$$[\mathbf{A}_N^H\mathbf{A}_N]_{r,q}(\alpha, \beta) = \begin{cases} \sum_{\kappa=0}^{MN-1} b_\kappa & \alpha = \beta \\ \sum_{\kappa=0}^{MN-1} \omega^{(\beta-\alpha)\kappa}b_\kappa & \text{otherwise} \end{cases}, \tag{2.141}$$

where, $\omega = e^{\frac{j2\pi}{N}}$, $b_\kappa = g((-rN+\kappa)_{MN}))g((-qN+\kappa)_{MN}))$, $r, q = 0\cdots M-1$ and $\alpha, \beta = 0\cdots N-1$. Let, $s = 0\cdots N-1$

$$[\mathbf{A}_N^H\mathbf{A}_N]_{r,q}((\alpha-s)_N, (\beta-s)_N) = \sum_{\kappa=0}^{MN-1} \omega^{(\beta-s-\alpha+s)\kappa}b_\kappa \quad = [\mathbf{A}_N^H\mathbf{A}_N]_{r,q}(\alpha, \beta) \tag{2.142}$$

This proves that each block $[\mathbf{A}_N^H\mathbf{A}_N]_{r,q}$ of $\mathbf{A}_N^H\mathbf{A}_N$ is circulant. Let, $\varsigma = 0, \cdots, M-1$.

$$[\mathbf{A}_N^H\mathbf{A}_N]_{(r-\varsigma)_M,(q-\varsigma)_M}(\alpha, \beta) = \sum_{\kappa=0}^{MN-1} \omega^{(\beta-\alpha)\kappa}g((-rN+\varsigma N+\kappa)_{MN})g((-qN+\varsigma N+\kappa)_{MN}). \tag{2.143}$$

As ω is periodic with N and $g(\beta)$ is periodic with MN, $[\mathbf{A}_N^H\mathbf{A}_N]_{(r-\varsigma)_M,(q-\varsigma)_M}(\alpha, \beta) = [\mathbf{A}_N^H\mathbf{A}_N]_{r,q}(\alpha, \beta)$, hence, $\mathbf{A}_N^H\mathbf{A}_N$ is block circulant matrix with circulant blocks(BCCB) with blocks of size $N \times N$. In the same way, it can be proved that $\mathbf{A}_M^H\mathbf{A}_M$ is also BCCB with blocks of size $M \times M$.

Using the properties of BCCB matrix, one gets that [80] (i) inverse of BCCB matrix is a BCCB matrix, (ii) addition of a diagonal matrix with equal elements to a BCCB matrix is a BCCB matrix, (iii) multiplication of two BCCB matrix is BCCB matrix. Hence, it can be easily proved that $\mathscr{B}$ and $\mathscr{C}\mathscr{C}^{\mathrm{H}}$ is also a BCCB in case of AWGN channel.

For any BCCB matrix [80] it can be shown that (i) diagonal elements are identical , (ii) all rows have equal power and (iii) all columns have equal power. Using this, it can be then concluded that, $P_{Sig,l} = P_{Sig}$, $P_{Sig+Inr,l} = P_{Sig+Inr}$, $P_{Inr,l} = P_{Inr}$ and $P_{Npp,l} = P_{Npp}$, $\forall l$. Therefore, we can write, $[\mathscr{B}]_{l,l} = \frac{trace\{\mathscr{B}\}}{MN}$, $[\mathscr{B}^2]_{l,l} = \frac{trace\{\mathscr{B}^2\}}{MN}$ and $[\mathscr{C}\mathscr{C}^{\mathrm{H}}]_{l,l} = \frac{trace\{\mathscr{C}\mathscr{C}^{\mathrm{H}}\}}{MN}$. Using this and (2.135), we can write,

$$P_{Sig} = \frac{\sigma_d^2}{(MN)^2} trace\{\mathscr{B}\}^2, \; P_{Sig+Inr} = \frac{\sigma_d^2}{MN} trace\{\mathscr{B}^2\} \text{ and } P_{Npp} = \frac{\sigma_\nu^2}{MN} trace\{\mathscr{C}\mathscr{C}^{H}\}. \tag{2.144}$$

Now, using (2.138) in above equation,

$$P_{Sig} = \frac{\sigma_d^2}{(MN)^2} | \sum_{s=0}^{MN-1} \frac{\lambda_s}{\lambda_s + \frac{\sigma_\nu^2}{\sigma_d^2}} |^2, \; P_{Sig+Inr}$$

$$= \frac{\sigma_d^2}{MN} \sum_{s=0}^{MN-1} | \frac{\lambda_s}{\lambda_s + \frac{\sigma_\nu^2}{\sigma_d^2}} |^2 \text{ and } P_{Npp} = \frac{\sigma_\nu^2}{MN} \sum_{s=0}^{MN-1} \frac{\lambda_s}{(\lambda_s + \frac{\sigma_\nu^2}{\sigma_d^2})^2}. \tag{2.145}$$

SINR can be computed as,

$$\gamma = \frac{P_{sig}}{P_{Inr+P_{Npp}}}. \tag{2.146}$$

Hence, SINR can be computed using the eigenvalues of $\mathbf{A}^H \mathbf{A}$. Through the above, because of the BCCB property, we can compute SINR easily than by using the direct form. Inversion of matrix $\frac{\sigma_\nu^2}{\sigma_d^2} \mathbf{I} + \mathbf{A}^H \mathbf{A}$ needs complexity of $O(M^3 N^3)$, whereas eigenvalue computation of $\mathbf{A}^H \mathbf{A}$ needs complexity of $O(NM \log_2 N)$ [80]. Hence, SINR computation using above method is much simpler than using direct matrix computation.

2.12.4.5 BER Computation
2.12.4.6 FSFC
Figure 2.22a shows cumulative distribution plot of signal plus interference value for 4 QAM modulated system for 3 dB and 9 dB $\frac{E_b}{N_0}$ values. CDF plot for both cases is compared with Gaussian CDF with measured mean and variance values. It is clear from the figure that interference plus noise values closely follow Gaussian distribution.

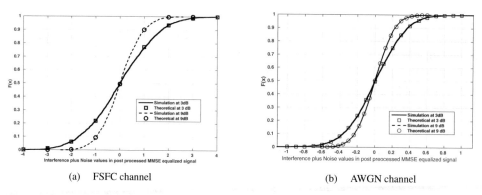

(a)   FSFC channel   (b)   AWGN channel

Figure 2.22   CDF plot of interference plus noise value in AWGN channel and FSFC for 4 QAM (pulse shaping filter is RRC with ROF =0.5).

Therefore, the BER for l^{th} QAM symbol with $\mathcal{M}$ modulation order can be computed as [81] by

$$P_b(E|\gamma_l) \simeq 4\frac{\sqrt{\mathcal{M}}-1}{\sqrt{\mathcal{M}}log_2(\mathcal{M})} \sum_{r=0}^{\sqrt{\mathcal{M}}/2-1} \left[Q\{(2r+1)\sqrt{\frac{3\gamma_l}{(\mathcal{M}-1)}}\} \right], \qquad (2.147)$$

γ_l is the post processing SINR for l^{th} symbol at the receiver given in (2.136). Average probability of error can be found as

$$P_b(E) = \frac{1}{MN} \times \sum_{l=0}^{MN-1} \int_0^\infty P_b(E|\gamma_l) f_{\gamma_l}(\gamma_l) d\gamma_l, \qquad (2.148)$$

where, $f_{\gamma_l}(\gamma_l)$ is probability density function of γ_l.

2.12.4.7 AWGN Channel
Figure 2.22b shows cumulative distribution plot of interference plus noise value (2.134) for 4 QAM modulated system for 3 dB and 9 dB $\frac{E_b}{N_0}$ values. As in case of FSFC, interference plus noise values closely follow Gaussian distribution. BER for QAM symbol with $\mathcal{M}$ modulation order can be computed as [81],

$$P_b(E) \simeq 4\frac{\sqrt{\mathcal{M}}-1}{\sqrt{\mathcal{M}}log_2(\mathcal{M})} \sum_{r=0}^{\sqrt{\mathcal{M}}/2-1} \left[Q\{(2r+1)\sqrt{\frac{3\gamma_l}{(\mathcal{M}-1)}}\} \right], \qquad (2.149)$$

where, γ is the post processing SINR given in (2.146).

2.12.4.8 Results
In this section, we test our developed BER expressions for MMSE receiver. GFDM system with parameters given Table 3.2 is considered here. It is assumed that the subcarrier

Table 2.4   Simulation parameters of GFDM receivers.

Number of Subcarriers N	128
Number of Timeslots M	5
Mapping	16 QAM
Pulse shape	RRC with ROF =0.1 or 0.5 or 0.9
CP length N_{CP}	16
Channel	AWGN and FSFC
Channel Length N_{ch}	16
Power delay profile	$[10^{-\frac{\alpha}{5}}]^{\mathrm{T}}$, where $\alpha = 0, 1 \cdots N_{ch} - 1$
Subcarrier Bandwidth	3.9 KHz
RMS delay Spread	4.3 μ sec
Coherence Bandwidth	4.7 KHz

bandwidth is larger than the coherence bandwidth of the channel for FSFC. SNR loss due to CP is also considered for FSFC.

BER vs $\frac{E_b}{N_o}$ for 16-QAM in AWGN and FSFC for MMSE receiver is presented in Figure 2.23. Legends "GFDM without self-interference" are obtained by using $\gamma_l = \frac{\sigma_d^2}{\sigma_\nu^2}$ in (2.148) for flat fading and in (2.149) for AWGN channel. This is a reference BER result for GFDM with zero self-interference and 16 QAM modulation in flat fading (2.148) and AWGN (2.149). The legends marked "Analytical" are obtained by first computing γ_l using (2.136) for FSFC and using (2.146) for AWGN and then using this γ_l in (2.148) and (2.149) to obtain average BER for FSFC and AWGN, respectively. The legends marked "Simu" are obtained using Monte Carlo simulations using parameters given in Table 2.5. It is seen that BER from simulation matches quite well with the analytical for both FSFC and AWGN. The gap between GFDM curve (analytical and simulation) and GFDM with no self-interference curve is due to self-interference encountered in GFDM [82]. From the above discussion, it can be concluded that the expressions developed in this work are useful in estimating the theoretical BER for MMSE based GFDM receiver in FSFC and AWGN channel.

2.12.5 Performance Comparison

In this section, we present the performance comparison of legacy waveforms. We have done Monte Carlo simulations to compare the performance of waveforms. For evaluation, we have considered the channel model, given in [74]. We have considered uncoded system and use of linear receiver structures as described in Chapter 2. The parameters of considered waveforms are provided in Table 2.5. It is assumed that the subcarrier bandwidth is comparable to the coherence bandwidth of the channel for FSFC. SNR loss due to CP is also considered for FSFC.

Figure 2.24 shows the comparison of BER vs $\frac{E_b}{N_0}$ of considered waveforms for 16 QAM modulation. UFMC performs better than vanilla forms of GFDM, FBMC and OFDM. FBMC performs better than OFDM at low SNR values and worse at high SNR values.

Next, we will see the effect of CFO on the BER performance of considered waveforms. We have considered five percent CFO value i.e., the frequency offset is five percent of

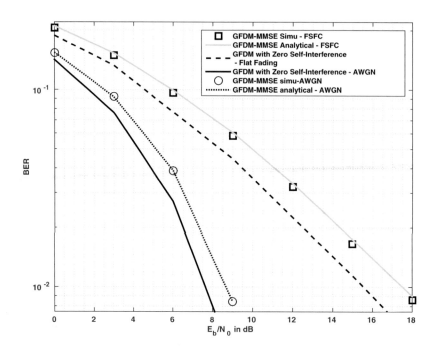

Figure 2.23   BER vs $\frac{E_b}{N_o}$ for MMSE GFDM receiver over AWGN and frequency selective channel for 16 QAM (RRC pulse shaping filter is used with ROF=0.5).

Table 2.5   Simulation parameters of comparison of waveforms.

Parameters	Attributes
Number of Subcarriers N	64
Number of Timeslots M (for GFDM)	5
Mapping	16 QAM
Pulse shape (GFDM)	RRC with ROF =0.1
Pulse shape (FBMC)	Phy-Dyas Filter [83]
Pulse shape (UFMC)	Equi-Ripple 120 dB attenuation FIR
Number of Resource Blocks (UFMC)	4
Block Based Pulse Shape for OFDM and GFDM	RRC with ROF = 0.35 and filter length = 17
CP length N_{CP}	16
Channel Length N_{ch}	16
Power delay profile	$[10^{-\frac{\alpha}{5}}]^{\mathrm{T}}$, where $\alpha = 0, 1 \cdots N_{ch} - 1$
Subcarrier Bandwidth	3.9 KHz
RMS delay Spread	4.3 μ sec
Coherence Bandwidth	4.7 KHz

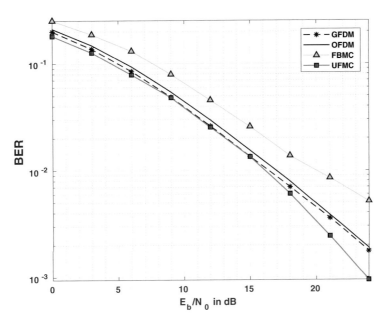

Figure 2.24   Comparison of BER for ideal channel and 16QAM modulation.

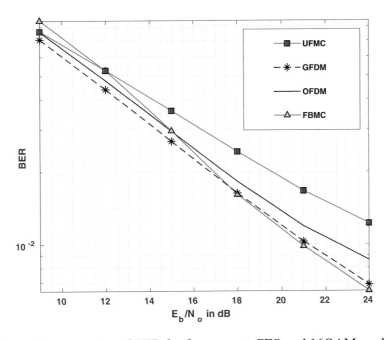

Figure 2.25   Comparison of BER for five percent CFO and 16QAM modulation.

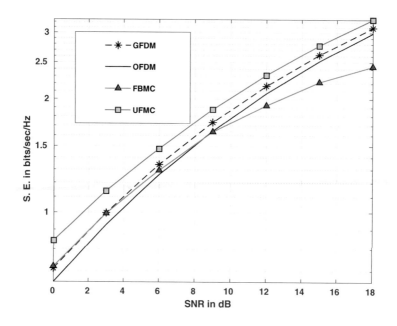

Figure 2.26   Comparison of Spectral Efficiency for Ideal Channel using Water Filling Algorithm.

subcarrier bandwidth. Figure 2.25 shows the comparison of BER vs $\frac{E_b}{N_0}$ of considered waveforms for 16 QAM modulation and five percent CFO. It is observed that FBMC outperforms all other waveforms in this case. GFDM performs better than OFDM, and UFMC. OFDM shows degradation of one order (10), whereas GFDM and FBMC demonstrate the deterioration of lesser than one order. FBMC and GFDM are rather resilient towards CFO due to the use of very well localised pulse shapes in the frequency domain. UFMC, which filters a group of subcarriers, is less resistant towards CFO than FBMC and GFDM.

Figure 2.26 shows the spectral efficiency of considered waveforms using optimum per symbol power derived from water-filling algorithm [42]. We assume perfect channel knowledge both at the transmitter as well as at receiver. It can be seen that UFMC has the highest spectral efficiency. GFDM has higher spectral efficiency than FBMC, OFDM. FBMC has higher spectral efficiency than OFDM and comparable capacity to GFDM at low SNR. At high SNR values, FBMC has lower spectral efficiency than OFDM. It can be concluded that SNR gain due to lesser CP length in GFDM manifests higher spectral efficiency in GFDM based waveforms than OFDM based waveforms. FBMC and UFMC waveforms have spectral efficiency gains because they do not use CP. However, in the case of FBMC, subcarriers become nonorthogonal in FSFC, which degrades its SINR, and hence, spectral efficiency.

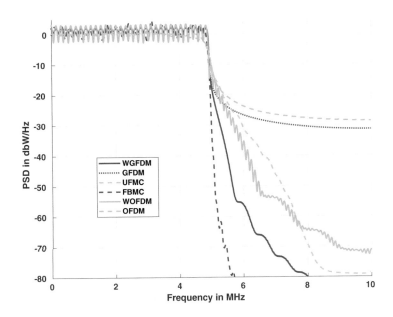

Figure 2.27 Comparison of out of band leakage.

Figure 2.27 shows one-sided power spectral density plot of considered waveforms. In a 20 MHz system, 128 sub-carriers are considered out of which 64 subcarriers are switched off (32 on each edge) to observe OoB characteristics of waveforms. Power spectral density is averaged over 10^4 transmitted symbols for each considered waveform. FBMC has the lowest stop band attenuation and narrowest transition band. UFMC has 58dB more stop-band attenuation than OFDM. UFMC has a very large transition band. GFDM has 5dB more stop band attenuation and narrower transition band than OFDM. GFDM has higher spectral leakage when compared with UFMC and FBMC. Reasons behind such high spectral leakage of GFDM are; (1) use of circular pulse shape and (2) use of CP. Circular pulse shape distorts the spectral properties specially at the edge subsymbols. CP introduces discontinuities in time domain which results in high spectral leakage. To reduce spectral leakage of GFDM and OFDM, a block based windowing can be used without any spectral efficiency loss and reliability loss [74]. Figure 2.27 depicts the spectral leakage performance of windowed GFDM (WGFDM) and windowed OFDM (WOFDM). It can be observed that WGFDM has 10 dB and 15 dB more stop-band attenuation than UFMC and WOFDM, respectively.

Finally, Figure 2.28 shows complementary cumulative distribution function (CCDF) plot of PAPR of all waveforms. 10^5 transmitted frames were generated, where each frame has four transmitted signal blocks. We compare the PAPR value that is exceeded with probability less than 0.01% ($\Pr\{\text{PAPR} > \text{PAPRo} = 10^{-4}\}$). UFMC has quite similar PAPR

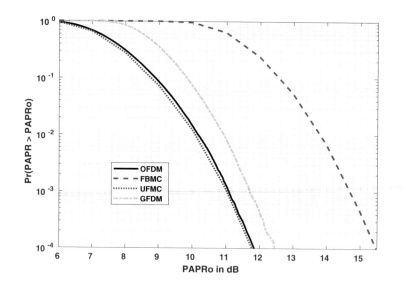

Figure 2.28 Comparison of PAPR.

to OFDM. GFDM has 0.7 dB, whereas FBMC, which has highest PAPR, has 3 dB worse PAPR than OFDM.

Based on the detail performance analysis of the contending waveforms, the following conclusions can be made.

In the case of CFO, FBMC and GFDM is performing the best. FBMC and GFDM achieve around 3 dB SNR gain over OFDM in the presence of CFO. UFMC provides highest spectral efficiency that is followed by GFDM, OFDM and lastly by FBMC at high SNR. FBMC has the lowest OoB leakage which is followed by windowed GFDM.

2.12.6 Issues with GFDM

2.12.6.1 High PAPR

PAPR of multi-carrier systems is high due to summations of multiple subcarriers. Pulse shaping further adds to the problem of PAPR [84]. Thus, GFDM suffers from high PAPR due to the use of multiple subcarriers and pulse shaping. This is depicted in Figure 2.28 which shows that GFDM has higher PAPR than OFDM. High PAPR is a critical problem especially for small modems such as handheld mobile devices, IOT devices etc. It is thus, vital to reduce the PAPR of GFDM which we tackle in Chapter 3.

2.12.6.2 High Computational Complexity

It is utmost important to compute the complexity of GFDM system as it is directly proportional to the cost of the system. Out of different mathematical operations, number of complex multiplication is a significant contributor to computational complexity [85]. Computational complexity is computed in terms of number of complex multiplications

Table 2.6 Number of complex multiplication of different receivers in GFDM.

Technique	Operations	Number of complex multiplications
GFDM Tx	one vector matrix multiplication for $\mathbf{y} = \mathbf{Ad}$	$(MN)^2$
GFDM Rx ZF-MF	1. Frequency domain equation (ZF) • one FFT operation for $\mathbf{y}' = \mathbf{Wy}$. • one diagonal complex valued matrix inversion and one diagonal matrix and vector multiplication for $\mathbf{y}'' = \mathbf{\Lambda}^{-1}\mathbf{y}'$. • one FFT operation for $\mathbf{y}_{FDE} = \mathbf{Wy}''$ 2. Matched filter for AWGN : one matrix vector multiplication for $\hat{d}_{ZF} = \mathbf{A}^H\mathbf{y}_{FDE}$	$\frac{3MN}{2}\log_2(MN)$ + $2MN + (MN)^2$
GFDM Rx ZF-ZF	1. Frequency domain equation (ZF): same operations as in ZF-MF. 2. ZF for AWGN : one matrix vector multiplication for $\hat{d}_{MF} = \mathbf{A}^{-1}\mathbf{y}_{FDE}$. Considering A^{-1} to be known and precomputed at the receiver.	$\frac{3MN}{2}\log_2(MN)$ + $2MN + (MN)^2$
GFDM Rx ZF-MMSE	1. Frequency domain equation (ZF): same operations as in ZF-MF. 2. MMSE for AWGN • MN complex value addition for computing $\mathbf{C} = \frac{\mathbf{I}}{\gamma} + \mathbf{A}^H\mathbf{A}$. • one matrix inversion for computing $\mathbf{C}^{-1}$. • one matrix vector multiplication for $\mathbf{y}_{temp} = \mathbf{A}^H\mathbf{y}$. • one matrix vector multiplication for $\hat{\mathbf{d}}_{MMSE} = \mathbf{C}^{-1}\mathbf{y}_{temp}$.	$\frac{3MN}{2}\log_2(MN)$ + $\frac{(MN)^3}{3} + 2(MN)^2$ + $\frac{2MN}{3}$
GFDM Rx ZF-SIC	1. Frequency Domain equation (ZF): same operations as in ZF-MF. 2. DSIC for AWGN • MF operation to compute $\mathbf{y}_{MF}$ and $MN(\sqrt{modorder} - 1)$ comparators for detection of matched filter $\hat{\mathbf{d}}_{detect}$. • $2M$ complex multiplication and $2M - 1$ complex addition are needed for each subcarrier and iteration index to compute $\mathbf{y}_k = \mathbf{Ad}_k$ where k is subcarrier index. • MN complex subtraction are needed for each subcarrier and iteration index to compute $\mathbf{y}_{interfree,k} = \mathbf{y}_{MF} - \mathbf{y}_k$	$\frac{3MN}{2}\log_2(MN)$ + $2MN + 2(MN)^2J$

required to implement transmitter and different receivers for GFDM system. It has been assumed that modulation matrix $\mathbf{A}$ is known at the receiver, hence, any matrix that is derived from $\mathbf{A}$ is also known to receiver. Minimum number of complex multiplications required to perform various matrix and vector operations is computed. Complexity computation can be found following Table 2.6.

2.13 Precoded GFDM System to Combat Inter Carrier Interference: Performance Analysis

In the previous section, we have seen that GFDM suffers from high PAPR. We explained that high PAPR in GFDM is due to its multi-carrier nature, further that pulse shaping adds to the problem of PAPR in GFDM. Such high PAPR is critical especially for devices having limited RF capabilities such as handheld mobile devices, IOT devices etc. In the previous section, we also explained that GFDM is a nonorthogonal transmission and suffers from self-interference. The signal gets further distorted when it goes through a wireless channel. This necessitates the use of high complexity advanced receiver. In this section , we describe precoding methods to reduce PAPR of GFDM systems. We also discuss channel aware precoding method which orthogonalizes GFDM transmission. Such precoding can reduce receiver signal processing complexity when channel is known a priory at the transmitter.

2.13.1 Section Contents

A detailed exposition of the product of the modulation matrix with its hermitian reveals some interesting properties which helps in developing precoding techniques. The details are described in this section. A detailed complexity analysis of different schemes is presented in this chapter which shows that D-SIC is not quite simple to implement in comparison with ZF receiver. To reduce the self-interference at the receiver, precoding techniques for GFDM are described. A generalized framework for precoding based GFDM is detailed. Based on the properties related to the modulation matrix as mentioned above, Block inverse discrete Fourier transform (BIDFT) based precoding is presented. Performance of BIDFT, discrete Fourier transform (DFT) and singular value decomposition (SVD) based precoding techniques are compared with uncoded GFDM. PAPR of precoded GFDM is compared with uncoded GFDM and OFDM as well. A summary of the main contents is given below.

1. Three new precoding techniques which improve performance of GFDM system are proposed.
2. We establish that SVD-based precoding for GFDM removes interference by orthogonalising the received symbols. When channel is constant for multiple transmit instances, the computational complexity of reception is reduced significantly.
3. We also propose block IDFT and DFT based precoding schemes which, without any increase in computational complexity, (1) reduces the PAPR significantly and (2) extract diversity gain in frequency selective fading channels.

Section outline

The first precoding scheme BIDFT is described in Section 2.13.2.1. The scheme is developed using special properties of $(\mathbf{HA})^{H}\mathbf{HA}$ (detailed in Section 2.13.2.2) and $\mathbf{A}^{H}\mathbf{A}$ (detailed in Section 2.12.4.3). For this precoding scheme there can be two kinds of receiver processing namely (i) joint processing : described in Section 2.13.2.2, which equalizes channel and GFDM modulation matrix simultaneously, whereas (ii) two-stage processing : described in Section 2.13.2.3, first stage equalizes for the channel and second stage equalizes for GFDM modulation matrix. Two types of precoding matrices are defined for BIDFT precoding namely (i) BIDFT-M : where $\mathbf{A}$ is structured in blocks of $M \times M$ i.e. $\mathbf{A}_M$ and (ii) BIDFT-N : where $\mathbf{A}$ is structured in blocks of $N \times N$ i.e. $\mathbf{A}_N$. BIDFT-N precoded GFDM is processed using joint processing as well as two-stage processing, whereas BIDFT-M precoded GFDM is processed using two-stage processing only.

DFT-based precoding is described in Section 2.13.2.4. SVD based precoding is described in Section 2.13.2.5. For each precoder scheme, Precoder matrix $\mathbf{P}$, corresponding receivers and post processing SNR is described in detail. Both BIDFT- and DFT-based precoding does not require channel state information (CSI) at the transmitter to compute $\mathbf{P}$, whereas SVD based precoding needs CSI at the transmitter to compute $\mathbf{P}$. Channel

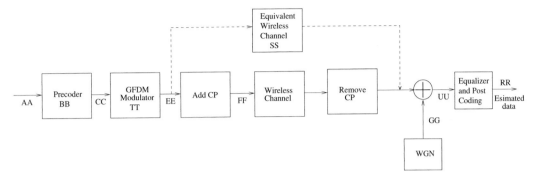

Figure 2.29   Transmitter and receiver architecture for precoding based GFDM system.

knowledge at the transmitter can be maintained via feedback from the receiver or through the reciprocity principle in a duplex system[86]. BER for precoded GFDM system is presented in Section 2.13.2.6. Computational complexity of GFDM and precoded GFDM system is given in Section 2.13.2.7.

2.13.2 Precoded GFDM System

Since, we consider precoding, let $\mathbf{P}$ be a precoding matrix of size $MN \times MN$, which is defined in Section 2.13.2. The data vector $\mathbf{d}$ be multiplied with precoding matrix $\mathbf{P}$ and we obtain precoded data vector $\tilde{\mathbf{d}} = \mathbf{P}\mathbf{d}$. The $MN \times 1$ precoded data vector $\tilde{\mathbf{d}} = [\tilde{d}_{0,0} \cdots \tilde{d}_{k,m} \cdots \tilde{d}_{N-1,M-1}]^{\mathrm{T}}$, where $k = 0 \cdots N - 1$ denote subcarrier index and $m = 0 \cdots M - 1$ indicates time slot index. Conventional GFDM system [74] can be seen as a special case of precoded GFDM system when, $\mathbf{P} = \mathbf{I}_{NM}$. The precoded data vector, $\tilde{\mathbf{d}}$, is modulated using GFDM Modulator. The flow of operations, as described below, can be understood in the light of Fig. 2.29.

2.13.2.1 Block IDFT Precoded GFDM
Received signal in (2.118) can be processed in two ways: (i) joint processing : channel and self interference are equalized simultaneously, and, (ii) two-staged processing: channel and self-interference and equalized separately.

2.13.2.2 Joint Processing
Suppose, the received signal in (2.118) passed through a matched filter. Equalized vector which can be given as,

$$\begin{aligned}
\mathbf{y}^{\mathrm{MF}} &= (\mathbf{HAP})^{\mathrm{H}}\mathbf{y} \\
&= \mathbf{P}^{\mathrm{H}}(\mathbf{HA})^{\mathrm{H}}\mathbf{HAPd} + (\mathbf{HA})^{\mathrm{H}}\boldsymbol{\nu}.
\end{aligned} \tag{2.150}$$

Since, $(\mathbf{HA})^{\mathrm{H}}\mathbf{HA}$ is multiplied to desired data vector in above equation. Now, we will explore some properties of $(\mathbf{HA})^{\mathrm{H}}\mathbf{HA}$

$(\mathbf{HA})^{\mathrm{H}}\mathbf{HA}$

Using, description of $\mathbf{A}_N$ given in section 2.12.4.3, $(\mathbf{HA}_N)^H\mathbf{HA}_N$ matrix can be given as,

$$(\mathbf{HA}_N)^H\mathbf{HA}_N = \begin{bmatrix} [\mathbf{L}_{0,0}]_{N\times N} & \cdots & [\mathbf{L}_{0,M-1}]_{N\times N} \\ \vdots & \ddots & \vdots \\ [\mathbf{L}_{M-1,0}]_{N\times N} & \cdots & [\mathbf{L}_{M-1,M-1}]_{N\times N} \end{bmatrix}_{MN\times MN}, \quad (2.151)$$

where, $\mathbf{L}_{u,v} = \mathbf{E}^H\mathscr{G}_u^H\boldsymbol{\Upsilon}\mathscr{G}_v\mathbf{E}$ is a $N \times N$ sub-matrix or block, where $\boldsymbol{\Upsilon} = \mathbf{H}^H\mathbf{H}$ is a $MN \times MN$ matrix and $u,v = 0\cdots M - 1$. Channel convolution matrix $\mathbf{H}$ is a circulant matrix and it can be shown that $\boldsymbol{\Upsilon} = \mathbf{H}^H\mathbf{H}$ is also a circulant matrix using the properties of circulant matrices[79]. $(\mathbf{HA}_N)^H\mathbf{HA}_N$ will be a block circulant matrix, iff, $\mathbf{L}_{(u+s)_M,(v+s)_M} = \mathbf{L}_{u,v}$, where $s = 0\cdots M - 1$. In the expression of $\mathbf{L}_{u,v}$, matrix $\mathbf{E}^H$ and $\mathbf{E}$ are independent of block indices u, v. Therefore, it can be said that, $\mathbf{G}$ is a block circulant matrix, if $\boldsymbol{\Phi}_{(u+s)_M,(v+s)_M} = \boldsymbol{\Phi}_{u,v}$, where, $\boldsymbol{\Phi}_{u,v} = \mathscr{G}_u^H\boldsymbol{\Upsilon}\mathscr{G}_v$ is $MN \times MN$ matrix. Let, $\boldsymbol{\Upsilon} = \{v_{r,q}\}_{MN\times MN}$ and using the definition of $\mathscr{G}_p$, it can be shown that

$$\boldsymbol{\Phi}_{u,v}(r,q) = g_{(-uN+r)_{MN}}g_{(-vN+q)_{MN}}v_{r,q} \text{ and} \quad (2.152)$$

$$\boldsymbol{\Phi}_{(u+s)_M,(v+s)_M}(r,q) = g_{(-(u+s)N+r)_{MN}}g_{(-(v+s)N+q)_{MN}}v_{r,q}. \quad (2.153)$$

Substituting, $r' = r - sN$ and $q' = q - sN$ and since, $\boldsymbol{\Upsilon}$ is a circulant matrix and hence $v_{(r'+sN,q'+sN)_{MN}} = v_{r',q'}$. Then, substituting $r' \& q'$ with $r \& q$,

$$\boldsymbol{\Phi}_{(u+s)_M,(v+s)_M}(r,q) = g_{(-uN+r)_{MN}}g_{(-vN+q)_{MN}}v_{r,q} = \boldsymbol{\Phi}_{(u)_M,(v)_M}(r,q) \quad (2.154)$$

Hence, $(\mathbf{HA}_N)^H\mathbf{HA}_N$ is a block circulant matrix with blocks of size $N \times N$. (It may be noted here that $(\mathbf{HA}_M)^H\mathbf{HA}_M$ will not be block circulant with blocks of size $M \times M$). Since, $(\mathbf{HA}_N)^H\mathbf{HA}_N$ is block circulant matrix with blocks of size $N \times N$, it can be decomposed as given in [80], as,

$(\mathbf{HA}_N)^H\mathbf{HA}_N = \mathbf{F}_{bN}\mathbf{D}_{bN}\mathbf{F}_{bN}^H$, where, $\mathbf{F}_{bN} = \begin{bmatrix} \mathbf{W}_N^0 & \mathbf{W}_N & \cdots \mathbf{W}_N^{M-1} \end{bmatrix}_{MN\times MN}$,

where, $\mathbf{W}_N^i = \dfrac{\begin{bmatrix} \mathbf{I}_N & w_N^i\mathbf{I}_N & \cdots & w_N^{i(N-1)}\mathbf{I}_N \end{bmatrix}^T_{N\times MN}}{\sqrt{N}}$, where, $w_N = e^{\frac{j2\pi}{N}}$ and $\mathbf{D}_{bN} = diag\{\mathbf{D}_{bN}^0\ \mathbf{D}_{bN}^1\cdots\mathbf{D}_{bN}^{M-1}\}$ is block diagonal matrix with blocks of size $N \times N$, where $\mathbf{D}_{bN}^r$ is r^{th} diagonal matrix of size $N \times N$.

Using this decomposition and taking $\mathbf{A}$ as $\mathbf{A}_N$, matched filter output in (2.150) can be written as,

$$\mathbf{y}^{MF} = \mathbf{P}^H\mathbf{F}_{bN}\mathbf{D}_{bN}\mathbf{F}_{bN}^H\mathbf{P}\mathbf{d} + (\mathbf{HA}_N)^H\boldsymbol{\nu}. \quad (2.155)$$

BIDFT-N precoding

if we choose, $\mathbf{P} = \mathbf{F}_{bN}$ (call it block inverse discrete Fourier transform -N (BIDFTN) precoding), then,

$$\mathbf{y}^{MF} = \mathbf{D}_{bN}\mathbf{d} + (\mathbf{H}\mathbf{A}_N)^{\mathrm{H}}\boldsymbol{\nu}$$

$$= \begin{bmatrix} \mathbf{D}_{bN}^0 & & & \\ & \mathbf{D}_{bN}^1 & & \\ & & \ddots & \\ & & & \mathbf{D}_{bN}^{N-1} \end{bmatrix} \begin{bmatrix} \mathbf{d}_0^N \\ \mathbf{d}_1^N \\ \vdots \\ \mathbf{d}_{M-1}^N \end{bmatrix} + \bar{\boldsymbol{\nu}}, \qquad (2.156)$$

where, $\bar{\boldsymbol{\nu}}$ is MF processed noise vector. In above equation $\mathbf{D}_{bN}$ being block diagonal matrix, adds only $N-1$ interfering symbols instead $MN-1$ (in case of uncoded GFDM). This shows that precoding reduces the number of interfering symbols significantly. Zero forcing equalization is applied to reduce interference further. Multiplying $\mathbf{D}_{bN}^{-1}$ in above equation we get,

$$\hat{\mathbf{d}}_{bidft}^{JP} = \mathbf{d} + \boldsymbol{\nu}_{bdft}^{JP}, \qquad (2.157)$$

where, $\boldsymbol{\nu}_{bdft}^{JP} = \mathbf{D}_{bN}^{-1}(\mathbf{H}\mathbf{A}_N)^{\mathrm{H}}\boldsymbol{\nu}$ is post processing noise vector and superscript JP signifies that signal processing steps followed in this method are joint processing (channel-and self-interference are equalized jointly). $\mathbf{D}_{bN}$ can be computed as,

$$\mathbf{D}_{bN} = \mathbf{F}_{bN}^{\mathrm{H}}(\mathbf{H}\mathbf{A}_N)^{\mathrm{H}}\mathbf{H}\mathbf{A}_N\mathbf{F}_{bN}. \qquad (2.158)$$

Post processing SNR for l^{th} symbol can be obtained as,

$$\gamma_{bidft,l}^{JP} = \frac{\sigma_{d^2}}{E[\boldsymbol{\nu}_{bdft}^{JP}(\boldsymbol{\nu}_{bdft}^{JP})^{\mathrm{H}}]_{l.l}}, \qquad (2.159)$$

where, denominator in above equation is post processing noise power for l^{th} symbol.

2.13.2.3 Two-Stage Processing

In above method, channel and GFDM were equalized together. In this method, we will first equalize channel distortions, and then GFDM induced self-interference. As explained in earlier, $\mathbf{H}$ is a circulant matrix. Hence, $\mathbf{H}$ can be decomposed as,

$$\mathbf{H} = \mathbf{W}_{NM}\boldsymbol{\Psi}\mathbf{W}_{NM}^{\mathrm{H}}, \qquad (2.160)$$

where, $\mathbf{W}_{NM}$ is normalized IDFT matrix of size $MN \times MN$ and $\boldsymbol{\Psi} = diag\{v_0, v_1, \cdots v_{MN-1}\}$ is a diagonal matrix. Channel equalized vector can be obtained as,

$$\mathbf{y}_{FDE} = \mathbf{W}_{NM}^{\mathrm{H}}\boldsymbol{\Psi}^{-1}\mathbf{W}_{NM}\mathbf{y} = \mathbf{A}\mathbf{P}\mathbf{d} + \mathbf{W}_{NM}\boldsymbol{\Psi}^{-1}\mathbf{W}_{NM}^{\mathrm{H}}\boldsymbol{\nu}, \qquad (2.161)$$

where, first term is the transmitted signal which is free from channel distortions completely, second term is enhanced noise and subscript FDE is acronym for frequency domain equalization (as above described channel equalization is frequency domain equalization

[43]). $\mathbf{\Psi}$ can be obtained as $\mathbf{\Psi} = \mathbf{W}_{NM}^{H}\mathbf{H}\mathbf{W}_{NM}$ which equivalently obtained by taking NM point FFT of zero padded channel convolution vector $\mathbf{h}$, which is also the first column of $\mathbf{H}$[43, 87].

Now, passing channel equalized data $\mathbf{y}_{FDE}$ to matched filter receiver, we can get,

$$\mathbf{y}_{FDE-MF} = (\mathbf{AP})^{H}\mathbf{y}_{FDE} = \mathbf{P}^{H}\mathbf{A}^{H}\mathbf{AP}\mathbf{d} + (\mathbf{AP})^{H}\mathbf{W}_{NM}\mathbf{\Psi}^{-1}\mathbf{W}_{NM}^{H}\boldsymbol{\nu}. \quad (2.162)$$

It has been proved in section 2.12.4.4 that $\mathbf{A}^{H}\mathbf{A}$ is BCCB matrix with blocks of size either $N \times N$ or $M \times M$, which depends on whether $\mathbf{A} = \mathbf{A}_N$ or $\mathbf{A} = \mathbf{A}_M$, which are described next.

BIDFT-N Precoding

When modulation matrix is defined as $\mathbf{A}_N$, $\mathbf{A}_N^{H}\mathbf{A}_N$ can be decomposed as,

$$\mathbf{A}_N^{H}\mathbf{A}_N = \mathbf{F}_{bN}\tilde{\mathbf{D}}_{bN}\mathbf{F}_{bN}^{H}, \quad (2.163)$$

where, $\tilde{\mathbf{D}}_{bN} = diag\{\mathbf{D}_{bN}^0, \mathbf{D}_{bN}^1 \cdots \mathbf{D}_{bN}^{M-1}\}$ is $MN \times MN$ block diagonal matrix where $\mathbf{D}_{bN}^r$ is r^{th} diagonal block of size $N \times N$. Choosing, $\mathbf{P} = \mathbf{F}_{bN}$ (BIDFTN precoding) and using above decomposition, matched filter output in (2.162) can be written as,

$$\mathbf{y}_{FDE-MF} = \tilde{\mathbf{D}}_{bN}\mathbf{d} + (\mathbf{A}\mathbf{F}_{bN})^{H}\mathbf{W}_{NM}\mathbf{\Psi}^{-1}\mathbf{W}_{NM}^{H}\boldsymbol{\nu}. \quad (2.164)$$

Now, multiplying $\tilde{\mathbf{D}}_{bN}^{-1}$ in above equation,

$$\hat{\mathbf{d}}_{FDE-MF-ZF}^{N} = \mathbf{d} + \boldsymbol{\nu}_{FDE-MF-ZF}^{N}, \quad (2.165)$$

where, $\boldsymbol{\nu}_{FDE-MF-ZF}^{N} = \tilde{\mathbf{D}}_{bN}^{-1}(\mathbf{A}\mathbf{F}_{bN})^{H}\mathbf{W}_{NM}\mathbf{\Psi}^{-1}\mathbf{W}_{NM}^{H}\boldsymbol{\nu}$ is enhanced noise vector. Since, $\tilde{\mathbf{D}}_{bN}$ needs to be computed to obtain $\hat{\mathbf{d}}_{FDE-MF_ZF}^{N}$, it can be computed as, $\tilde{\mathbf{D}}_{bN} = \mathbf{F}_{bN}\mathbf{A}^{H}\mathbf{A}\mathbf{F}_{bN}^{H}$. Post processing SNR for l^{th} symbol can be obtained as,

$$\gamma_{FDE-MF-ZF,l}^{N} = \frac{\sigma_{d^2}}{E[\boldsymbol{\nu}_{FDE-MF-ZF}^{N}(\boldsymbol{\nu}_{FDE-MF-ZF}^{N})^{H}]_{l,l}}, \quad (2.166)$$

where, denominator in above equation is enhanced noise power for l^{th} symbol.

BIDFT-M precoding

Now, if modulation matrix is defined as $\mathbf{A}_M$, $\mathbf{A}_M^{H}\mathbf{A}_M = \mathbf{F}_{bM}\mathbf{D}_{bM}\mathbf{F}_{bM}^{H}$, where, $\mathbf{F}_{bM} = \begin{bmatrix} \mathbf{W}_M^0 & \mathbf{W}_M & \cdots & \mathbf{W}_M^{N-1} \end{bmatrix}_{MN \times MN}$, where, $\mathbf{W}_M^r = \frac{\begin{bmatrix} \mathbf{I}_M & w_M^r\mathbf{I}_M & \cdots & w_M^{r(M-1)}\mathbf{I}_M \end{bmatrix}^T}{\sqrt{M}}_{M \times MN}$, where, $w_M = e^{\frac{j2\pi}{M}}$ and $\mathbf{D}_{bM} = diag\{\mathbf{D}_{bM}^0 \ \mathbf{D}_{bM}^1 \cdots \mathbf{D}_{bM}^{N-1}\}$ is block diagonal matrix with blocks of size $M \times M$, where $\mathbf{D}_{bM}^r$ is r^{th} diagonal matrix of size $M \times M$. Using this decomposition, choosing $\mathbf{P} = \mathbf{F}_{bM}$ (block IDFT-M (BIDFTM) Precoding) and following same signal processing steps as in case of $\mathbf{A}_N$, equalized data vector can be given as,

$$\hat{\mathbf{d}}_{FDE-MF_ZF}^{M} = \mathbf{d} + \boldsymbol{\nu}_{FDE-MF-ZF}^{M}, \quad (2.167)$$

where, $\nu^M_{FDE-MF-ZF} = \mathbf{D}^{-1}_{bM}(\mathbf{AF}_{bM})^H \mathbf{W}_{NM} \mathbf{\Psi}^{-1} \mathbf{W}^H_{NM} \nu$ is enhanced noise vector. Since, $\mathbf{D}_{bM}$ needs to be computed to obtain $\hat{\mathbf{d}}^M_{FDE-MF_ZF}$, it can be computed as,

$$\mathbf{D}_{bM} = \mathbf{F}_{bM}\mathbf{A}^H \mathbf{AF}^H_{bM}. \tag{2.168}$$

Post processing SNR for l^{th} symbol can be obtained as,

$$\gamma^M_{FDE-MF-ZF,l} = \frac{\sigma_{d^2}}{E[\nu^M_{FDE-MF-ZF}(\nu^M_{FDE-MF-ZF})^H]_{l,l}}, \tag{2.169}$$

where, denominator in above equation is enhanced noise power for l^{th} symbol.

In summary, BIDFT precoding can be understood by Fig. 2.30a.

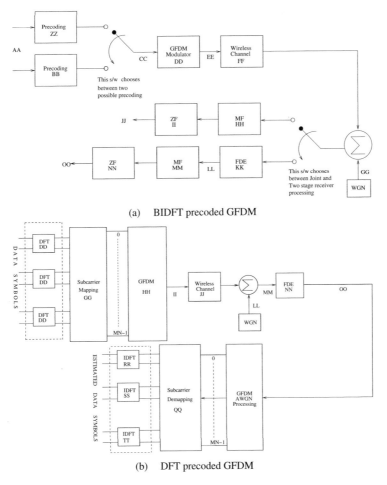

(a)   BIDFT precoded GFDM

(b)   DFT precoded GFDM

Figure 2.30   Block diagram of BIDFT and DFT precoded GFDM system.

2.13.2.4 DFT Precoded GFDM

DFT precoding has been used in OFDM systems[43, 87]. It has been shown that DFT precoding reduces PAPR significantly and is one of the optimum precoding matrices to reduce PAPR in OFDM systems [88]. This motivates us to investigate DFT precoded GFDM for reducing PAPR in GFDM system. Suppose, Q is spreading factor of the system then DFT order/ size can be computed as, $N_{DFT} = \frac{N}{Q}$. We assume that Q divides N completely. Two subcarrier mapping schemes are considered in this work: [89] (i) localized frequency division multiple access (LFDMA) and (ii) interleaved frequency division multiple access (IFDMA).

Precoding matrix $\mathbf{P}$ can be defined as $\mathbf{P} = \mathbf{P}_m \mathbf{P}_c$, where $\mathbf{P}_c$ is a block diagonal matrix with each block being a DFT spreading matrix and $\mathbf{P}_m$ is a permutation matrix which implements subcarrier mapping i.e., LFDMA or IFDMA. The precoding spreading matrix $\mathbf{P}_c$ can be written as,

$$
\mathbf{P}_c = \begin{bmatrix} \mathbf{W}_{N_{DFT}} & & & \\ & \mathbf{W}_{N_{DFT}} & & \\ & & \ddots & \\ & & & \mathbf{W}_{N_{DFT}} \end{bmatrix}, \tag{2.170}
$$

where, $\mathbf{W}_{N_{DFT}}$ is normalized DFT matrix of size $N_{DFT} \times N_{DFT}$. Permutation matrix, $\mathbf{P}_m$ for LFDMA is an identity matrix. DFT precoded GFDM system can be understood from Fig. 2.30b. Precoded data vector is GFDM modulated using the modulation matrix $\mathbf{A}$. Received signal can be equalized using conventional linear[65, 74] or nonlinear[68] equalizer. We will present here ZF receiver for DFT precoded GFDM. ZF equalized precoded data vector can be obtained as,

$$
\hat{\tilde{\mathbf{d}}}_{zf} = (\mathbf{HA})^{-1}\mathbf{y} \ = \tilde{\mathbf{d}} + (\mathbf{HA})^{-1}\boldsymbol{\nu}. \tag{2.171}
$$

Equalized data vector $\hat{\mathbf{d}}_{dft-spread}$ can be obtained as,

$$
\hat{\mathbf{d}}_{dft-spread} = \mathbf{P}^H \hat{\tilde{\mathbf{d}}}_{zf} \ = \underbrace{\mathbf{P}_c^H}_{De-spreading} \times \underbrace{\mathbf{P}_m^H}_{Subcarrier\,De-mapping} \times \hat{\tilde{\mathbf{d}}}_{zf} = \mathbf{d} + \boldsymbol{\nu}_{dft-zf},
$$
$$\tag{2.172}$$

where, $\boldsymbol{\nu}_{dft-zf} = \mathbf{P}^H((\mathbf{HA}))^{-1}\boldsymbol{\nu}$ is post processing noise vector. Post processing SNR for l^{th} symbol can be obtained as,

$$
\gamma_{dft-ZF,l}^N = \frac{\sigma_{d^2}}{E[\boldsymbol{\nu}_{dft-zf}^N (\boldsymbol{\nu}_{dft-zf}^N)^H]_{l,l}}, \tag{2.173}
$$

where, denominator in above equation is enhanced noise power for l^{th} symbol.

2.13.2.5 SVD Precoded GFDM

The product of the channel matrix and the modulation matrix ($\mathbf{HA}$) can be decomposed as,

$$
\mathbf{HA} = \mathbf{USV}^H, \tag{2.174}
$$

where, $\mathbf{U}$ and $\mathbf{V}$ are unitary matrix and $\mathbf{S}$ is diagonal singular-value matrix i.e., $\mathbf{S} = diag\{s_0, \ s_1 \cdots s_r, \cdots s_{MN-1}\}$, where s_r is r^{th} singular value. Then, $\mathbf{y}$ in (2.118), can be written as,

$$\mathbf{y} = \mathbf{U}\mathbf{S}\mathbf{V}^H\mathbf{P}\mathbf{d} + \boldsymbol{\nu}. \tag{2.175}$$

At transmitter, by choosing $\mathbf{P} = \mathbf{V}$ (assuming ideal feedback channel) and multiplying both sides with $\mathbf{U}^H$, the estimated symbol vector can be written as,

$$\hat{\mathbf{d}}^{svd} = \mathbf{S}\mathbf{d} + \mathbf{U}^H\boldsymbol{\nu}. \tag{2.176}$$

The estimated l^{th} symbol is then given by,

$$\hat{d}_l^{svd} = s_l d_l + \sum_{q=0}^{MN-1} [\mathbf{U}^H]_{l,q}\nu(q). \tag{2.177}$$

From the above, it can be seen that by using SVD-based precoding, interference is completely removed by orthogonalization of $\mathbf{HA}$ without the need for matrix inversion, which is required in zero forcing (ZF) and minimum mean square error (MMSE) receiver. SINR for l^{th} symbol, can be computed as,

$$\gamma_l^{svd} = \frac{\sigma_d^2}{\sigma_\nu^2}|s_l|^2. \tag{2.178}$$

2.13.2.6 BER Performance of Precoding Techniques

Expression for estimated data symbols, given in(2.157,2.165,2.167,2.172,2.177), are summations of desired data symbols and enhanced noise vector. The enhanced noise vector is weighted sum of complex Gaussian random variable for a given channel realization. Hence, enhanced noise is also complex Gaussian random vector for a given channel realization. BER for QAM symbol with modulation order $\mathcal{M}$ over FSFC can be obtained as,

$$P_b(E|\gamma_l) \simeq 4\frac{\sqrt{\mathcal{M}} - 1}{\sqrt{\mathcal{M}}\log_2(\mathcal{M})} \sum_{r=0}^{\sqrt{\mathcal{M}}/2-1} \left[Q\{(2r+1)\sqrt{\frac{3\gamma_l}{(\mathcal{M}-1)}}\} \right], \tag{2.179}$$

where, γ_l is post processing SNR computed in (2.159,2.166,2.169,2.173,2.178). Average probability of error can be found as,

$$P_b(E) = \frac{1}{MN} \times \sum_{l=0}^{MN-1} \int_0^\infty P_b(E|\gamma_l) f_{\gamma_l}(\gamma_l) d\gamma_l, \tag{2.180}$$

where, $f_{\gamma_l}(\gamma_l)$ is probability distribution function of SINR for l^{th} symbol.

2.13.2.7 Computational Complexity

Precoding schemes for GFDM are proposed in previous subsections. Computation complexity computed in terms of number of complex multiplications required to implement

Table 2.7   Number of complex multiplications of different techniques in GFDM.

Technique	Operations	Number of Complex Multiplications
SVD precoded GFDM Tx	two vector matrix multiplications for computing $\mathbf{AVd}$	$2(MN)^2$
SVD precoded GFDM Rx (Known SVD)	one vector matrix multiplication for computing $\mathbf{U}^H\mathbf{y}$	$(MN)^2$
SVD precoded GFDM Rx (Un-Known SVD)	1. SVD computation 2. one vector matrix multiplication for computing $\mathbf{U}^H\mathbf{y}$	$(MN)^2 + 26(MN)^3$
BIDFT precoded GFDM Tx	one vector matrix multiplication for computing $\mathbf{AF}_b\mathbf{d}$ as $\mathbf{AF}_b$ can be precomputed at transmitter	$(MN)^2$
BIDFT precoded GFDM Rx (Joint Processing)	1. one vector matrix multiplication to compute $\mathbf{HAF}_b$ 2. Computation of block diagonal matrix $\mathbf{D}_b$. 3. Inversion $\mathbf{D}_b^{-1}$ which can be computed by inverting M square matrices of order N.	$(MN)^2 \log_2(N) + 2(MN)^2 + MN^2$

BIDFT precoded GFDM Rx (ZF-ZF)	1. Frequency domain equalization : Same as in ZF-MF.
	2. AWGN processing :
	  • one vector matrix multiplication for computing $\mathbf{y}_{MF} = (\mathbf{AF}_b)^H\mathbf{y}$.
	  • Computation of block diagonal matrix $\mathbf{D}_b^{-1}$.
	  • N times square matrix inversion of order M and block diagonal matrix and vector multiplications to compute $\hat{\mathbf{d}}_{BDFT} = \mathbf{D}_b^{-1}\mathbf{y}_{MF}$
		$\frac{3MN}{2}\log_2(MN) + 2MN + NM^2 + (MN)^2 + \underbrace{NM^2}_{\text{BIDFTM}} \text{ or } \underbrace{MN^2}_{\text{BIDFTN}}$

DFT Precoded GFDM Tx (additional over GFDM Tx)	1. additionally, MQ times N_{DFT} point FFT and subcarrier mapping 2. same operation as in GFDM Tx	$\frac{MN}{2}\log_2 N_{DFT}$
DFT Precoded GFDM Rx (additional over GFDM Rx)	1. same operation as in GFDM Rx. 2. additionally, MQ times N_{DFT} point FFT and subcarrier mapping 3. same operation as in GFDM Tx	$\frac{MN}{2}\log_2 N_{DFT}$

transmitter and receiver of precoded and uncoded GFDM system are presented here. It is considered that that modulation matrix $\mathbf{A}$ is known at the receiver, hence, any matrix that is derived from $\mathbf{A}$ is also known to receiver, such as, $\mathbf{A}^H$, $\mathbf{A}^{-1}$, etc. All known receivers for uncoded GFDM are considered for complexity computation of uncoded GFDM and DFT precoded GFDM. The minimum number of complex multiplications required to perform various matrix and vector operations is computed. Complexity of SVD precoded GFDM receiver is computed for two cases namely: (i)SVD of channel is (useful when channel is static for multiple transmit instances) and (ii) SVD of channel is unknown. Complexity computation can be found following Table 2.7.

2.13.3 Results

In this section, the results related to works described in earlier sections are presented. Analytical evaluation of BER of MMSE receiver with GFDM is given in Section 2.12.4.

Table 2.8 Simulation parameters of GFDM receivers.

Number of subcarriers N	128
Number of timeslots M	5
Mapping	16 QAM
Pulse shape	RRC with ROF =0.1 or 0.5 or 0.9
CP length N_{CP}	16
Channel	AWGN and FSFC
Channel length N_{ch}	16
Power delay profile	$[10^{-\frac{\alpha}{5}}]^{\mathrm{T}}$, where $\alpha = 0, 1 \cdots N_{ch} - 1$
Subcarrier bandwidth	3.9 KHz
RMS delay spread	4.3 μ sec
Coherence bandwidth	4.7 KHz

BER evaluation of precoded GFDM system is provided in Section 2.13.2.6. Complexity of different transmitters and receivers of GFDM and precoded GFDM is given in Section 2.13.3.2. Finally, PAPR of proposed precoding schemes is compared with GFDM and OFDM in Section 2.13.3.3. GFDM system with parameters given Table 2.7 is considered here. It is assumed that the subcarrier bandwidth is larger than the coherence bandwidth of the channel for FSFC. SNR loss due to CP is also considered for FSFC.

2.13.3.1 BER Evaluation of Precoded Techniques

BER of precoded GFDM system is evaluated via Monte–Carlo simulations. Simulation parameters are given in Table 3.2. One thousand channel realizations are used to obtain The BER results. the legends GFDM-ZF, GFDM-MMSE and GFDM-DSIC are used to represent the performance of the corresponding receivers for GFDM. The legends SVD-Prec, BIDFTN, BIDFTM, BIDFT-JP, LFDMA-ZF and IFDMA-ZF, are used to indicate the result of SVD-based precoding, block IDFTN precoding with two-stage processing, block IDFTM with two-stage processing, block IDFTN with joint processing, DFT precoding with IFDMA and DFT precoding with LFDMA, respectively. Legend OFDM-CP is used to indicate theoretical BER performance of OFDM-CP. Fig. 2.31 shows BER vs $\frac{E_b}{N_o}$ for ROF = 0.1 in AWGN channel and FSFC. The following observations can be made. Under AWGN, all schemes have similar performance. It may be noted that in AWGN, there is no SNR loss as CP is not required. The performance of GFDM is only slightly worse than OFDM. In the case of FSFC, it is seen that SVD, BIDFT-N based precoding has performance similar to OFDM, as ICI is low because of small ROF. The better performance of DFT and BIDFT-M precoded GFDM over OFDM can be attributed to frequency diversity gain, which can be understood from Figure 2.32. In this figure, it can be seen that one symbol in BIDFT-M has larger frequency spread compared to IFDMA which is larger than LFDMA. Accordingly, BIDFTM precoded GFDM is performing better than IFDMA precoded GFDM which is performing better than LFDMA precoded GFDM.

Next, we look at Figure 2.33 which shows similar performance analysis for ROF 0.9, which indicates higher ICI. In case of AWGN, degradation can be observed as compared

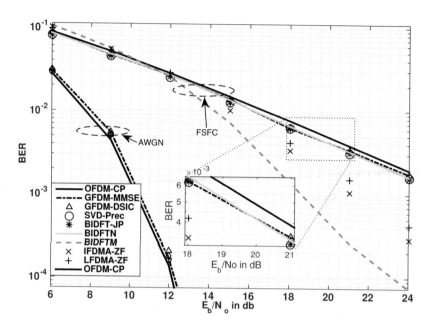

Figure 2.31   BER vs $\frac{E_b}{N_o}$ for precoded GFDM receiver over frequency selective and AWGN channel using 16-QAM with N=128, M=5, ROF=0.1 (RRC).

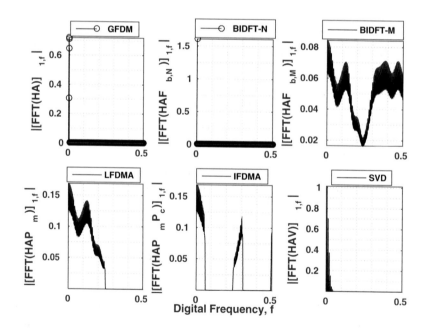

Figure 2.32   Frequency spread of one symbol for different transmission schemes.

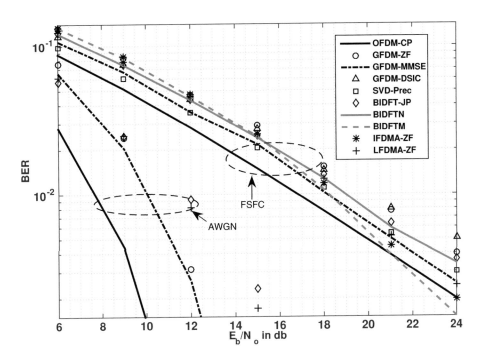

Figure 2.33 BER vs $\frac{E_b}{N_o}$ for precoded GFDM receiver over frequency selective and AWGN channel using 16-QAM with N=128,M=5, ROF=0.9 (RRC).

to Figure 2.31. We see a similar degradation for all precoding schemes in FSFC. This is expected for ROF 0.9 as significant overlapping of pulses in frequency.

It is clear that the performance of precoding schemes is sensitive to ROF. In this work we have only considered RRC pulse shape. It is shown in [90, 74] that performance of uncoded GFDM is sensitive to pulse shape choice. Proposed precoded GFDM system will also be sensitive to pulse shape choice as the enhanced noise expressions in (36, 43, 46, 51, and 56) are function of modulation matrix **A** which is a function of pulse shape as given in (7).

It can be concluded that precoding schemes are giving good performance over FSFC. BIDFTM precoded GFDM performs best among all precoding schemes.

2.13.3.2 Complexity Computation

Table 2.10 compares the complexity of different transmitters and Table 2.11 compares the complexity of different receivers for two different application scenarios given in [74]. Table 2.9 shows values of N and M considered for different application scenarios. Complexity is computed in terms of number of complex multiplications. Complexities of precoded GFDM systems and uncoded GFDM systems are compared with OFDM system. Spreading factor, $Q = 4$, is considered for DFT precoded GFDM. GFDM receiver

Table 2.9   Values of N and M for different application scenarios [74].

	Tactile internet(TI)	Wireless RAN (WRAN)
(N, M)	(128, 5)	(16, 127)

Table 2.10   Number of complex multiplications for different transmitters.

CASE	OFDM	Uncoded GFDM	SVD precoded GFDM	BIDFT precoded GFDM	DFT precoded GFDM
TI	4.4×10^3	4.09×10^5	8.19×10^5	4.09×10^5	4.12×10^5
WRAN	8.1×10^3	4.1×10^5	8.25×10^5	4.12×10^5	4.13×10^5

Table 2.11   Number of complex multiplications for different receivers.

CASE	OFDM	GFDM-ZF	GFDM-MMSE	GFDM-DSIC	SVD-GFDM Rx (known SVD)	SVD-GFDM Rx (unknown SVD)	BIDFT-GFDM joint processing	BIDFT-GFDM two-stage processing	DFT-GFDM
TI	8.9×10^3	4.19×10^5	8.8×10^7	3.2×10^6	4.09×10^5	6.8×10^9	3.7×10^6	5×10^5	4.2×10^5
WRAN	1.62×10^4	4.16×10^6	2.8×10^9	3.3×10^7	4.1×10^6	2.1×10^{11}	2.4×10^7	4.19×10^6	4.16×10^6

processing for DFT precoded GFDM is considered to be GFDM-ZF. Number of iterations for DSIC receiver is considered to be 4[68]. It is observed that the complexity of uncoded GFDM transmitter is about 100 times higher than OFDM transmitter in case of TI and, around 50 times greater than OFDM transmitter in case of WRAN. BIDFT precoded and DFT precoded GFDM transmitter has the same order of complexity as uncoded GFDM transmitter. Complexity of SVD precoded GFDM is around two times higher than uncoded GFDM transmitter. Therefore, it can be concluded that there is no significant increase in complexity for BIDFT and DFT precoded transmitter when compared with uncoded GFDM transmitter. However, for SVD precoded transmitter complexity is doubled, which is also not a significant increment.

GFDM-ZF receiver complexity is much higher than OFDM receiver for instance 50 times in TI scenario. SVD precoded GFDM receiver when SVD is known, two-stage BIDFT precoded receiver and DFT precoded receiver have around same complexity as GFDM-ZF receiver. GFDM-MMSE receiver has a very high complexity; for example it 2000 times higher than GFDM-ZF in TI scenario. BIDFT precoded receiver complexity, when SVD of channel is not known, is even higher, for instance, it is 100 times higher than the complexity of GFDM-MMSE. It is also important to note that GFDM-DSIC has higher complexity than GFDM-ZF, DFT precoded GFDM and two-stage BIDFT precoded GFDM.

2.13.3.3 PAPR of Precoding Techniques

The impact of precoding on PAPR is presented in Fig. 2.34. Complementary cumulative distribution function (CCDF) of PAPR is computed using Monte–Carlo simulation. 10^5 transmitted blocks were generated, where each block has two precoded GFDM symbols.

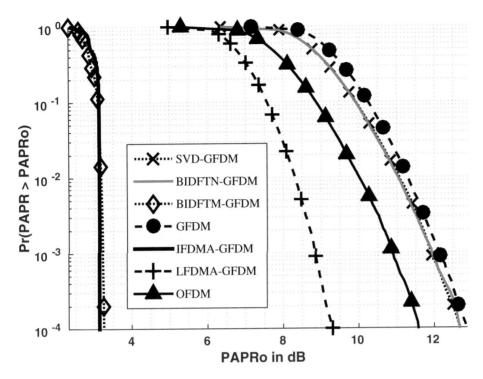

Figure 2.34 Complementary cumulative distribution function (empirical) of PAPR for GFDM for N=128 and M=5, ROF =0.5 (RRC) and Q= 4.

For each precoded GFDM symbol : $N = 128$, $M = 5$ and ROF for RRC pulse shaping filter = 0.5. For OFDM, $N = 128$. Randomly generated data symbols are considered to be QPSK modulated.

As expected, GFDM has worse performance than OFDM. Precoding has a positive effect on GFDM. We compare the PAPR value that is exceeded with probability less than 0.1% (Pr{PAPR > PAPRo = 10^{-3}}). SVD precoded GFDM and BIDFTN precoded GFDM reduces the PAPR by 0.3 dB, but it is still higher than OFDM. DFT precoded GFDM with LFDMA subcarrier mapping reduces PAPR by 3.4 dB and is lower than OFDM. DFT precoded GFDM with IFDMA subcarrier mapping and BIDFTM precoded GFDM reduces PAPR by 9 dB.

2.14 Chapter Summary

In this chapter, a survey of waveform design for transmission in doubly dispersive channels has been presented. This survey reveals the fundamental performance limit of waveforms under LTV channels. Works available in the literature discuss design of pulse shape in the TF domain to minimize the effect of the doubly dispersive channels. Broadly, it is seen that a

TF localized pulse is an approximate eigenfunction of doubly dispersive channels. However, arbitrary TF localization is not possible due to the TF uncertainty principle, which means that a perfect orthogonal transmission in a doubly dispersive channel is not possible, which motivated researchers to minimize the effects of channel. In this regard, typical waveform design steps for doubly dispersive channels when the second order of channel statistics are known are studied. Such approaches are divided into two steps: (i) coarse optimization; wherein TF lattice parameters are adjusted to minimize overall interference and (ii) pulse shaped design; wherein pulse shape is designed whose ambiguity function matches with channel scattering function.

A summary of legacy waveforms such as OFDM, W-OFDM, F-OFDM, FBMC, and UFMC is also included. The discussion brings out the advantages and disadvantages of these waveforms, which helps one to choose appropriate waveforms for future generations. GFDM is also introduced in this chapter. Important constraints of GFDM system, which are hindrance to its practical realizability, namely: (i) high PAPR and (ii) high computational complexity are also highlighted.

Three precoding techniques which improve performance of GFDM are also described. The SVD-based precoding for GFDM removes interference by orthogonalizing the received symbols. It does not require matrix inversion, yet its performance is quite close to that of GFDM-MMSE. The complexity of the receiver, for such precoding is quite high when compared to GFDM-ZF receiver when SVD of channel is not known. However, when SVD of channel is known, complexity of the receiver is comparable to GFDM-ZF. Hence, SVD-based precoding is found to be usable in cases when channel is nearly unchanged for multiple transmit instances. Performance of BIDFTN precoding is found to be better than GFDM-ZF and GFDM-DSIC. Two-stage BIDFTN precoded GFDM has complexity similar to GFDM-ZF and lesser than GFD M-DSIC as well as gives lower PAPR. Hence, two-stage BIDFTN precoded GFDM is found preferable over GFDM-ZF. BIDFTM precoded as well as DFT precoded GFDM receivers which require complexity similar to GFDM-ZF performs much better than even GFDM-MMSE receiver under FSFC. Apart from this, BIDFTM and DFT precoded GFDM reduces PAPR significantly. Precoded GFDM system described above is shown to have better BER performance than GFDM system without any increase in complexity with the added advantage of decreased PAPR. It can be concluded that BIDFTM precoded GFDM is preferable over other precoding schemes as it gives better BER performance than other precoded and decreases PAPR significantly without any increase in complexity.

3

OTFS Signal Model

This chapter describes the basic system model of OTFS which lays the foundation of the following chapters.

3.1 Introduction

In this chapter we present OTFS as another multi-carrier modulation scheme, which has evolved over time. With this approach we directly take a headlong dive into the system model for OTFS. We will describe the different aspects and views of OTFS as they appear with the flow. As an opening sentence, we can view OTFS as a two-dimensional spreading applied on QAM symbols, which are then modulated using OFDM or generalized frequency division multiplexing (GFDM) as shown in later chapters.

In OTFS, the user data (constellation symbols) is placed in the delay-Doppler (De-Do) domain as opposed to TF grid in OFDM. The data is then spread across the TF grid using a unitary transform. This is followed by an OFDM [91] or block OFDM [92] modulator. Cyclic prefix (CP) [91] or block CP [92] is added to absorb the channel delay spread. When OFDM modulation is used with CP, it is called as CP-OTFS whereas, OTFS with block OFDM modulation and block CP is termed as reduced CP OTFS (RCP-OTFS) [93]. In this chapter, we consider the RCP-OTFS, which is spectrally more efficient than CP-OTFS.

Notation

We use the following notations throughout the chapter. We let $\mathbf{x}$, $\mathbf{X}$ and x represent vectors, matrices, and scalars, respectively. The superscripts $(-)^{\mathrm{T}}$ and $(-)^{\dagger}$ indicate transpose and conjugate transpose operations, respectively. Notations $\mathbf{0}$, $\mathbf{I}_N$ and $\mathbf{W}_L$ represent zero matrix, identity matrix with order N and L-order normalized inverse discrete Fourier transform (IDFT) matrix respectively. Kronecker product operator is given by $\otimes$. The operator $\mathrm{diag}\{\mathbf{x}\}$ creates a diagonal matrix with the elements of vector $\mathbf{x}$. Circulant matrix is represented by $circ\{\mathbf{x}\}$ whose first column is $\mathbf{x}$. Notations $E\{-\}$ and $\lceil - \rceil$ are expectation and ceil operators, respectively. Column-wise vectorization of matrix $(\mathbf{X})$ is represented by

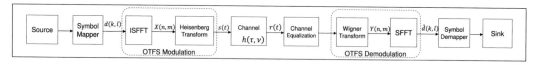

Figure 3.1   Block diagram of OTFS system.

$vec\{\mathbf{X}\}$. Natural numbers are denoted by $\mathbb{N}$. Complex conjugate value of x is given by $\bar{x}$ whereas $j = \sqrt{-1}$.

We consider an OTFS system with M number of subcarriers having Δf subcarrier bandwidth and N number of symbols having T symbol duration. Total bandwidth $B = M\Delta f$ and total duration $T_f = NT$. Moreover, OTFS system is critically sampled i.e., $T\Delta f = 1$. The system model description is summarized in the Figure 3.1.

3.2 OTFS Signal Generation

QAM modulated data symbols, $d(k, l) \in \mathbb{C}$, $k \in \mathbb{N}[0\ N-1]$, $l \in \mathbb{N}[0\ M-1]$, are arranged over Do-De lattice $\Lambda = \{(\frac{k}{NT}, \frac{l}{M\Delta f})\}$. We assume that $E[d(k,l)\bar{d}(k',l')] = \sigma_d^2 \delta(k - k', l - l')$, where δ is Dirac delta function. Do-De domain data $d(k,l)$ is mapped to TF domain data $X(n,m)$ on lattice $\Lambda^\perp = \{(nT,\ m\Delta f)\}$, $n \in \mathbb{N}[0\ N-1]$ and $m \in \mathbb{N}[0\ M-1]$ by using inverse symplectic fast Fourier transform (ISFFT). $X(n,m)$ can be given as [91],

$$X(n,m) = \frac{1}{\sqrt{NM}} \sum_{k=0}^{N-1} \sum_{m=0}^{M-1} d(k,l)e^{j2\pi[\frac{nk}{N} - \frac{ml}{M}]}. \tag{3.1}$$

Next, $X(n,m)$ is converted to a time domain signal $s(t)$ through a Heisenberg transform as,

$$s(t) = \sum_{n=0}^{N-1} \sum_{m=0}^{M-1} X(n,m)g(t-nT)e^{j2\pi m\Delta f(t-nT)}, \tag{3.2}$$

where, $g(t)$ is transmitter pulse of duration T. It has been shown in [92] that nonrectangular pulse induces nonorthogonality which degrades BER performance. Thus, here we consider a rectangular pulse i.e.

$$g(t) = \begin{cases} 1\ , \text{if } 0 \le t \le T \\ 0\ , \text{otherwise.} \end{cases} \tag{3.3}$$

To obtain discrete time representation of OTFS transmission, following the figures 3.1 and 3.2 , $s(t)$ is sampled at the sampling interval of $\frac{T}{M}$ [92]. Samples of $s(t)$ in $\mathbf{s} =$

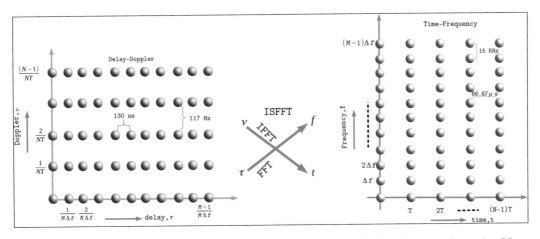

Figure 3.2 Do-De to TF conversion using inverse symplectic Fourier transform for $N = 128$ and $M = 512$.

$[s(0)\ s(1) \cdots s(MN-1)]$ and QAM symbols $d(k,l)$ are arranged in $M \times N$ matrix as,

$$\mathbf{D} = \begin{bmatrix} d(0,0) & d(1,0) & \cdots & d(N-1,0) \\ d(0,1) & d(1,1) & \cdots & d(N-1,1) \\ \vdots & \vdots & \ddots & \vdots \\ d(M-1,0) & d(M-1,1) & \cdots & d(N-1,M-1) \end{bmatrix}. \tag{3.4}$$

Using above formulations, $\mathbf{s}$ can be given as [92],

$$\mathbf{s} = vec\{\mathbf{D}\mathbf{W}_N\}. \tag{3.5}$$

Alternatively, if $\mathbf{d} = vec\{\mathbf{D}\}$, then the transmitted signal can also be written as matrix-vector multiplication,

$$\mathbf{s} = \mathbf{A}\mathbf{d}, \tag{3.6}$$

where, $\mathbf{A} = \mathbf{W}_N \otimes \mathbf{I}_M$ is the OTFS modulation matrix.

3.3 RCP-OTFS as Block OFDM with Time Interleaving

If $\mathbf{D}$ denotes the Do-De domain data symbols in matrix form, defined as ,

$$\mathbf{D} = \begin{bmatrix} d(0,0) & d(0,1) & \cdots & d(0,N-1) \\ d(1,0) & d(1,1) & \cdots & d(1,N-1) \\ \vdots & \vdots & \ddots & \vdots \\ d(M-1,0) & d(M-1,1) & \cdots & d(M-1,N-1) \end{bmatrix} \tag{3.7}$$

Then, the same time domain signal s can be obtained as,

$$s = vec\{\mathbf{W_M Z}\} \tag{3.8}$$

where, $\mathbf{Z}$ is the TF domain signal in matrix form given as,

$$\mathbf{Z} = \mathbf{W_M}^\dagger \mathbf{D} \mathbf{W_N} \tag{3.9}$$

which is simplified as,

$$s = vec\{\mathbf{D} \mathbf{W_N}\} \tag{3.10}$$

If the the same data,$\mathbf{X}$ as used in OFTS, is transmitted using block OFDM [1] then, $\mathbf{X}$ is placed in TF domain with N subcarriers and M OFDM symbols. In such a case, the time domain signal for block OFDM can be expressed as,

$$s_{\text{bofdm}} = vec\{\mathbf{W_N} \mathbf{D}^T\}. \tag{3.11}$$

From (3.10) and (3.11),

$$s(nM + m) = s_{\text{bofdm}}(mN + n) \tag{3.12}$$

where, $m \in \mathbb{N}[0 \ M - 1]$ and $n \in [0 \ N - 1]$.

OTFS as Block OFDM

Observing 3.12, one can understand that the RCP OTFS signal s can be obtained by interleaving block OFDM signal s_{bofdm}. ***Thus, it can be stated that RCP-OTFS can be seen as block OFDM with time interleaving.***

However, it is important to keep in mind that the parmeters for OTFS and its equivalent block OFDM system are different. For the same TF resource of T_f sec and B Hz, OTFS system has M subcarriers and N OTFS symbols, whereas equivalent block OFDM system has $M_{bofdm} = N$ subcarriers and $N_{bofdm} = M$ OFDM symbols. Therefore, the subcarrier bandwidth for the equivalent Block OFDM sytem changes to $\Delta f_{bofdm} = \frac{M}{N} \Delta f$ and OFDM symbol duration changes to $T_{bofdm} = \frac{N}{M} T$. In general $M > N$, thus $\Delta f_{bofdm} > \Delta f$ which implies that OTFS will have increased capability to combat Doppler as opposed to an OFDM system having Δf subcarrier bandwidth. Figure 3.3 shows the time domain signal generation for different waveforms and depicts the RCP-OTFS as block OFDM with time interleaving. The signal flow of an OFDM transmitter and that of a RCP OTFS transmitter are also shown in Figure 3.4a and Figure 3.4b, respectively.

3.4 Performance in AWGN Channel

3.4.1 Receiver for AWGN

The received signal after passing through an AWGN channel can be written as,

$$\mathbf{r} = \mathbf{s} + \mathbf{n}, \tag{3.13}$$

---

[1]In block OFDM, a block of OFDM symbols is given just one cyclic prefix unlike typical OFDM systems where each OFDM symbol uses its own cyclic prefix.

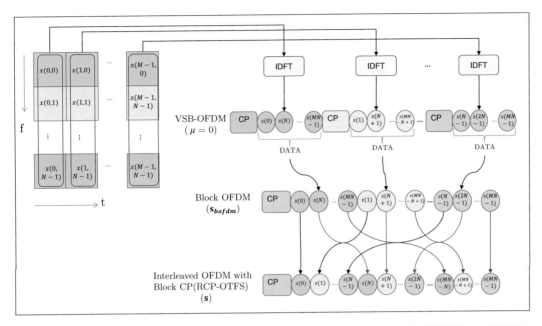

Figure 3.3 Time domain signal construction in CP-OFDM, Block OFDM and RCP-OTFS.

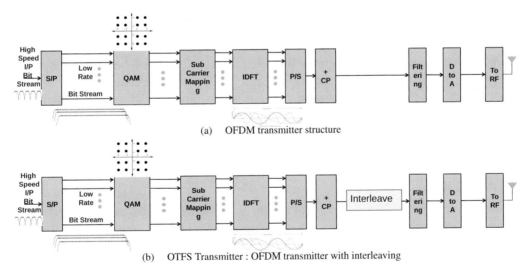

(a)   OFDM transmitter structure

(b)   OTFS Transmitter : OFDM transmitter with interleaving

Figure 3.4 Similarity of OTFS transmitter with that of an OFDM transmitter.

where, $\mathbf{n}$ is white Gaussian noise vector of length MN with elemental variance $\sigma_{\mathbf{n}}^2$. To recover the data, a matched filter (MF) is employed at the receiver, which can be given as,

$$\hat{\mathbf{d}} = \mathbf{A}^{\dagger}\mathbf{r}. \tag{3.14}$$

Table 3.1   Simulation parameters.

Parameter	Value
Carrier frequency(f_c)	6 GHz
Bandwidth(B)	7.68 MHz
Frame time(T_f)	10 ms
Subcarrier bandwidth(Δf)	15 KHz
OTFS parameters	M=512, N=128
OFDM parameters	M=512, N=128
Single carrier bandwidth(B)	7.68 MHz

As can be observed in (3.6), $\mathbf{A}$ is a unitary matrix. Using this the above equation is further simplified as,

$$\hat{\mathbf{d}} = \underbrace{\mathbf{A}^\dagger \mathbf{A}}_{\mathbf{I}} \mathbf{d} + \mathbf{A}^\dagger \mathbf{n} = \mathbf{d} + \tilde{\mathbf{n}}. \qquad (3.15)$$

It can be observed that recovered signal is free from interference. Also, using the unitary property of $\mathbf{A}$, it can be easily verified that $E[\tilde{\mathbf{n}}\tilde{\mathbf{n}}^\dagger] = \sigma_n^2 \mathbf{I}$. Thus, postprocessing noise $\tilde{\mathbf{n}}$ is uncorrelated which means that the BER performance of OTFS in AWGN channel is supposed to be similar to the performance of single carrier systems in AWGN channel.

3.4.2 Ber Performance in AWGN

Now, we present the BER performance of OTFS, OFDM, and single carrier system. The simulation parameters are mentioned in Table 3.1. BER performance of different waveforms under AWGN channel and 4 QAM mapping is plotted in figure 3.5. It can be observed that all three waveforms, namely single carrier, OTFS, and OFDM, have the same performance in AWGN channel. These results confirm the unitary transformations of data symbols in OTFS as well as in OFDM signals.

3.5 Performance in Time Varying Wireless Channel

3.5.1 The Channel

We consider a time varying channel of P paths with h_p complex attenuation, τ_p delay and ν_p Doppler value for pth path where $p \in [1\ P]$. Thus, Do-De channel spreading function can be given as,

$$h(\tau, \nu) = \sum_{p=1}^{P} h_p \delta(\tau - \tau_p) \delta(\nu - \nu_p) \qquad (3.16)$$

The delay and Doppler values for pth path are given as $\tau_p = \frac{l_p}{M\Delta f}$ and $\nu_p = \frac{k_p}{NT}$, where $l_p \in \mathbb{N}[0\ M-1]$ and $k_p \in \mathbb{N}[0\ N-1]$ are delay and Doppler bin number on Do-De lattice Λ for p^{th} path. We assume that N and M are sufficiently large, so that there is no effect of fractional delay and Doppler on the performance. Let τ_{max} and ν_{max} be the maximum delay and Doppler spread. Channel delay length $l_\tau = \lceil \tau_{max} M\Delta f \rceil$ and channel Doppler

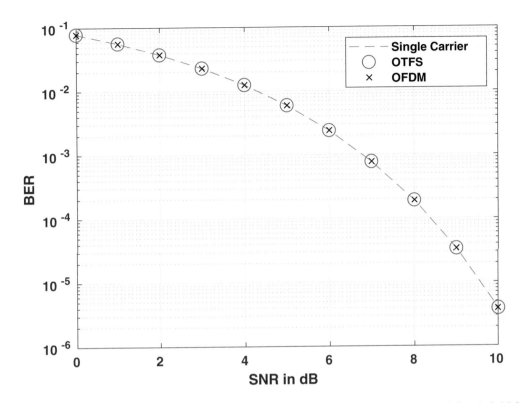

Figure 3.5 BER performance of different waveforms under AWGN channel for 4-QAM mapping.

length, $k_\nu = \lceil \nu_{max} NT \rceil$. Typically, $l_\tau << MN$ as well as $k_\nu << MN$ which dictates the system matrices to be sparse (as will be seen in Sec. 4.2.2.1).

We also define time varying frequency response of channel $h_{tf}(t, f)$ as,

$$h_{tf}(f, t) = \int_0^{\tau_{max}} \int_0^{\nu_{max}} h(\tau, \nu) e^{j2\pi(\nu t - f\tau)} d\nu d\tau \qquad (3.17)$$

which simplifies as,

$$h_{tf}(f, t) = \sum_{p=1}^{P} h_p e^{j2\pi(\nu_p t - f\tau_p)} \qquad (3.18)$$

Its discrete version $\acute{h}(m, n)$, the channel coefficient at mth subcarrier of nth time symbol used later in 6.1.5.2, is defined as,

$$\acute{h}(m, n) = h_{tf}(f, t)\Big|_{f=m\Delta f, t=nT} \qquad (3.19)$$

3.5.2 Linear Receivers

After removal of CP at the receiver, the received signal can be written as [94],

$$\mathbf{r} = \mathbf{Hs} + \mathbf{n} \tag{3.20}$$

where, $\mathbf{n}$ is white Gaussian noise vector of length MN with elemental variance $\sigma_{\mathbf{n}}^2$ and $\mathbf{H}$ is a $MN \times MN$ channel matrix and can be given as,

$$\mathbf{H} = \sum_{p=1}^{P} h_p \Pi^{l_p} \Delta^{k_p} \tag{3.21}$$

with $\Pi = circ\{[0 \quad 1 \quad 0 \cdots 0]_{MN \times 1}^T\}$ is a circulant delay matrix and $\Delta = diag\{[1 \quad e^{j2\pi\frac{1}{MN}} \cdots e^{j2\pi\frac{MN-1}{MN}}]^T\}$ is a diagonal Doppler matrix. The expression 3.21 can be derived as given in Appendix A.

3.5.2.1 MMSE Equalization

Here, the Do-De data is estimated by equalizing the received time domain signal using MMSE crietria [94] as,

$$\hat{\mathbf{d}} = \mathbf{H_{mmse}}\mathbf{r}, \tag{3.22}$$

where,

$$\mathbf{H_{mmse}} = (\hat{\mathbf{H}}\mathbf{A})^{\dagger}[(\hat{\mathbf{H}}\mathbf{A})(\hat{\mathbf{H}}\mathbf{A})^{\dagger} + \frac{1}{\Gamma}\mathbf{I}]^{-1}.$$

Therefore, the estimated data obtained after LMMSE equalization can be written as,

$$\hat{\mathbf{d}} = (\mathbf{HA})^{\dagger}[(\mathbf{HA})(\mathbf{HA})^{\dagger} + \frac{\sigma_{\nu}^2}{\sigma_d^2}\mathbf{I}]^{-1}\mathbf{r}. \tag{3.23}$$

$$\hat{\mathbf{d}} = \mathbf{A}^{\dagger}\mathbf{H}^{\dagger}[\mathbf{HH}^{\dagger} + \frac{\sigma_{\nu}^2}{\sigma_d^2}\mathbf{I}]^{-1}\mathbf{r}. \tag{3.24}$$

The MMSE matrix formed in (3.24) is a large order square matrix as MN is usually large. The direct implementation of LMMSE would require a complex multiplications in order of $O(M^3N^3)$. The MMSE receiver realized above is defined on the received time domain signal. The channel matrix is a time domain channel matrix. Thus, it may be noted that the equalization is agnostic to the modulation. Therefore, it can be applied to both OFDM as well as OTFS equivalently. A signal flow of the receiver for OFDM and OTFS with time domain equalization be seen in Figure 3.6. More detailed description is provided in Section 4.2.2.4 and Section 6.2.7.

BER results

Here, we present BER performance of the in EVA channel. Simulation parameters are given in Table 3.2.The CP is chosen long enough to accommodate the wireless channel delay spread. We plot BER for different waveforms in Figure 3.7 and Figure 3.8 for 500 Kmph and 30 Kmph vehicular speed respectively. Direct implementation of our receivers are obtained by using (3.24).

Two important observations can be made from Figure 3.7 and Figure 3.8.

1. OTFS-LMMSE receiver has 10dB gain over OFDM-MMSE receiver at BER of 10^{-3}. Since both OTFS and OFDM use identical time domain equalization which equalizes ICI due to Doppler, it can be said that OTFS can extract diversity gain.
2. The use of time domain MMSE receiver makes both the systems (block OFDM and RCP-OTFS) virtually invariant to Doppler, i.e., OTFS has unchanged performance at both high and low mobility conditions.

3.5.2.2 ZF Receiver for TVMC

After describing the LMMSE receiver we continue to state the other popular linear receiver viz., zero forcing (ZF) structure for OTFS.

It is very well established that the performance of zero forcing (ZF) receiver is inferior to MMSE receiver due to larger noise enhancement in ZF compared to MMSE. For example, ZF is shown to perform inferior to MMSE in case of CDMA systems in [95]. Similar observations are discussed in [96] for MIMO systems.

We denote ZF matrix for unitary $\mathbf{A}$, as,

$$\mathscr{C}_{zf} = \mathbf{A}^\dagger H^\dagger \underbrace{[HH^\dagger]^{-1}}_{\mathcal{B}_{zf}}. \tag{3.25}$$

Whereas, for MMSE $\mathscr{C}_{mmse} = \mathbf{A}^\dagger H^\dagger \underbrace{[HH^\dagger + \frac{\sigma_v^2}{\sigma_d^2}\mathbf{I}]^{-1}}_{\mathcal{B}_{mmse}}$, where $\mathbf{A}$ and $\mathbf{H}$ are OTFS

modulation matrix and channel matrix, respectively, and $\dagger$ denotes hermitian operation. Linear receiver's performance is affected by noise enhancement. For ZF and MMSE operation, $\mathcal{B}_{zf}$ and $\mathcal{B}_{mmse}$ need to be inverted. Thus, noise enhancement factor (NEF) for ZF and MMSE receiver depends on condition number of $\mathcal{B}_{zf}$ and $\mathcal{B}_{mmse}$, respectively. NEF (χ) for linear receivers whose receiver matrix is $\mathcal{B}$ can be defined as,

$$\chi = \frac{max[b_0, b_1 \cdots b_{MN-1}]}{min[b_0, b_1 \cdots b_{MN-1}]}, \tag{3.26}$$

where, $[b_0 \ b_1 \cdots b_{MN-1}]$ are eigen values of $\mathcal{B}$. We denote NEF for ZF and MMSE receivers as χ_{zf} and χ_{mmse} respectively. In case of additive white Gaussian noise channel (AWGN),

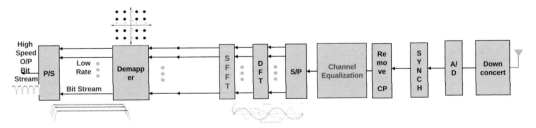

Figure 3.6 Time domain MMSE equalization for OTFS and OFDM.

Table 3.2   Simulation parameters.

Number of subcarriers M	512
Number of time slots N	128
Mapping	4 QAM
Subcarrier bandwidth	15 KHz
Channel	EVA [30]
Vehicular speed	500 Kmph
Carrier frequency	4 GHz

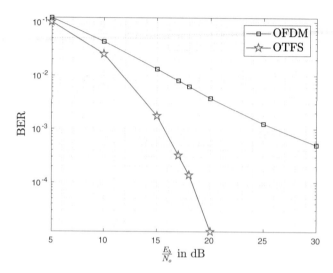

Figure 3.7   The BER performance of different waveforms for $v = 500$ Kmph.

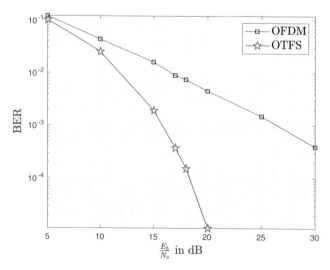

Figure 3.8   The BER performance of different waveforms for $v = 30$ Kmph.

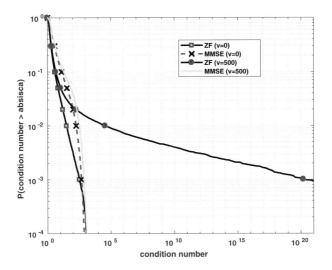

Figure 3.9 CCDF plot for condition number of ZF and MMSE matrices for 30 dB SNR, $N = 32$, $M = 32$ and vehicular speed of 500 Kmph.

$\mathbf{H} = \mathbf{I}_{MN}$, and in turn χ_{zf} as well as χ_{mmse} becomes unity which means there is no noise enhancement. When, $\mathbf{H}$ is a unitary operator, χ_{zf} and χ_{mmse} also becomes unity. But, in general wireless channel is nonorthogonal ($\chi > 1$) which results in noise enhancement. As condition number of $\mathbf{H}$ increases, the difference between the post processing SNR of ZF and MMSE also increases [97].

As, χ is a function of channel matrix $\mathbf{H}$, it is a random variable. We plot complementary cumulative distribution function (CCDF) of χ_{zf} and χ_{mmse} in Figure 3.9 for 30 dB SNR with $N = 32$, $M = 32$. We compute χ_{zf} and χ_{mmse} for 4×10^4 independent channel generations using power delay profile of EVA channel for vehicular speed of 0 Kmph and 500 Kmph.

Doppler is generated using Jake's formula, $\nu_p = \nu_{max} cos(\theta_p)$, where θ_p is uniformly distributed over $[-\pi \ \pi]$. It can be observed that NEF of ZF receiver is similar to that of MMSE receiver when vehicular speed is zero. When vehicular speed is 500 Kmph, NEF for ZF receiver is far greater than that of MMSE receiver. NEF of ZF can go as high as 10^{20}, whereas for NEF of MMSE is limited to 10^5. This indicates far worse performance of ZF in comparison to MMSE when vehicular velocity is 500 Kmph.

NEF of ZF receiver increases further for practical numerology of OTFS. For instance, we provide CCDF plot of NEF of ZF and MMSE receiver for $N = 128$ and $M = 512$ in Figure 3.10 and 3.11 for vehicular speed of 0 Kmph and 500 Kmph respectively. For zero vehicular speed, the NEF for ZF receiver goes upto 10^5 whereas for MMSE receiver it is limited to around 10^4. When velocity is further increased to 500 Kmph, with 90 percent probability NEF of ZF receievr is more than 10^{50}. In contrast, NEF for MMSE receiver has value lesser than 2×10^4 with 90 percent probability.

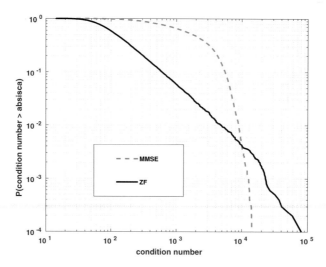

Figure 3.10 CCDF plot for condition number of ZF and MMSE matrices for 30 dB SNR, $N = 128$, $M = 512$ and vehicular speed of 0 Kmph.

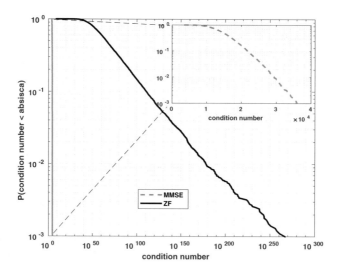

Figure 3.11 CCDF plot for condition number of ZF and MMSE matrices for 30 dB SNR, $N = 128$, $M = 512$ and vehicular speed of 500 Kmph.

3.5.2.3 BER Evaluation of ZF and MMSE Receiver

We present BER performance of ZF and MMSE receivers in EVA channel using Monte–Carlo simulations in Figure 3.12 for N=M=32 at vehicular speeds of 0 and 500 Kmph. It can be observed that BER of ZF receiver is floored at the value of 8×10^{-3} when vehicular speed is 500 Kmph. MMSE receiver has about 9 dB gain over ZF receiver at the BER value of 8×10^{-3}.

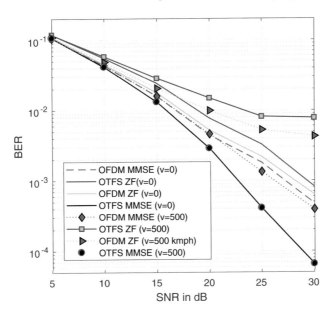

Figure 3.12 BER performance of ZF and MMSE matrices for $N = 32$ and $M = 32$, vehicular speed of 0 and 500 Kmph.

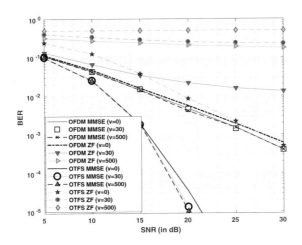

Figure 3.13 BER performance of ZF and MMSE matrices for $N = 128$ and $M = 512$ and vehicular speed of 0, 30, and 500 Kmph.

Figure 3.13 shows the BER vs SNR curves for ZF and MMSE receiver with $N = 128$ and $M = 512$. When vehicular speed is zero , OTFS-ZF receiver for has 13 dB SNR loss at the BER of 6×10^{-6} compared to OTFS-MMSE receiver. At 30 Kmph vehicular speed, ZF-OTFS floors at the BER of 10^{-2}. When speed is further increased to 500 Kmph, ZF receiver due to very high NEF, has BER values of 0.5 at each SNR point.

Thus, the following conclusions can be drawn from the above observations.

1. ZF receiver for OTFS is unable to extract diversity gain.
2. Interference at the receiver output is very high due to very high noise enhancement, especially at high vehicular speed.

3.6 Chapter Summary

In this chapter, the system model for OTFS has been described. It is shown that reduced CP OTFS signals can be generated by interleaving the signal of a block OFDM system. The performance of OTFS in AWGN channel is shown, where it is found to be identical with single carrier as well as OFDM. Thereafter, the linear time varying channel is described. It is used in the performance analysis of linear receivers for OTFS. The performance of MMSE and ZF receivers for OTFS is described in the chapter. Relative performance of OTFS and OFDM is also included to serve as a reference.

4

Receivers Structures for OTFS

The direct form of linear receivers for OTFS are described in Chapter 3. These structures have exorbitant complexity. It was also presented that the ZF receiver for OTFS has worse performance than MMSE. Therefore, it is no longer considered for further elaboration. In this chapter we describe three different receiver structure, for OTFS which have realization structural complexity. The first one of them is the belief propagation (BP) receiver described in Section 4.1. A low complexity version of the MMSE receiver is explained in Section 4.2. An interative successive interference cancellation receiver based on the LMMSE structure is given in Section 4.3.

4.1 Belief Propagation Receiver for a Sparse Systems

In this section, we describe a BP algorithm for equalization of channel in a sparse system. The BP algorithms are used for approximate maximum apposterior probability (AMAP) decoding which yields good bit error rate (BER) performance. Approximation of MAP occurs due to the assumption of independence between received samples, which is discussed below. This assumption holds to a certain extent for communication system, where channel matrices are sparse. OTFS channel matrix is sparse as will be elaborated later. Hence, we describe the BP algorithm for general sparse systems. In this chapter, we use belief propagation and message passing (MP) interchangably as both describe the same algorithm in the contex of this book.

4.1.1 Maximum Apposterior Probability (MAP) Decoding

If received signal vector $\mathbf{y}$ can be written as,

$$\mathbf{y} = \mathbf{Hx} + \mathbf{v} \tag{4.1}$$

where, $\mathbf{x}$ is the vector of M transmitted constellation symbols, i.e., $\mathbf{x}(i) \in \mathbb{A}$ and $\mathbb{A}$ is the set of constellation symbols. $\mathbf{H}$ is the $N \times M$ channel matrix. Then, the decoded data $\hat{\mathbf{x}}$

using maximum aposterior probability (MAP) decoding can be given as,

$$\hat{\mathbf{x}} = \underset{x \in \mathbb{A}^{M \times 1}}{\operatorname{argmax}} Pr(\mathbf{x}|\mathbf{y}, \mathbf{H}) \tag{4.2}$$

The above decoding of $\hat{\mathbf{x}}$ using MAP will be in order of exponential in M which is very high. Hence, we apply MAP procedure elementwise as,

$$\hat{\mathbf{x}}(c) = \underset{a_j \in \mathbb{A}}{\operatorname{argmax}} Pr(\mathbf{x}(c) = a_j|\mathbf{y}, \mathbf{H}), c = 1, 2, \cdots, M. \tag{4.3}$$

Using Baye's theorem,

$$\hat{\mathbf{x}}(c) = \underset{a_j \in A}{\operatorname{argmax}} Pr(\mathbf{y}|\mathbf{x}(c) = a_j, \mathbf{H}) Pr(\mathbf{x}(c) = a_j) \tag{4.4}$$

We assume that the symbols are equiprobable, hence,

$$\hat{\mathbf{x}}(c) = \underset{a_j \in A}{\operatorname{argmax}} Pr(\mathbf{y}|\mathbf{x}(c) = a_j, \mathbf{H}) \frac{1}{Q}$$

Further, we define sets I_d and J_c as $I_d = c|\mathbf{H}(d,c) \neq 0$ and $J_c = c|\mathbf{H}(d,c) \neq 0$. Therefore,

$$\hat{\mathbf{x}}(c) \approx \underset{a_j \in A}{\operatorname{argmax}} \prod_{d \in J_c} Pr(\mathbf{y}(d)|\mathbf{x}(c) = a_j, \mathbf{H})$$

The above expression is valid only when received elements $\mathbf{y}(d)$(s) are independent, which is hardly observed in a wireless channel. Therefore, belief propagation algorithm gives approximate MAP solutions. These approximate solutions are close to exact solutions, when the channel matrix $\mathbf{H}$ is sparse in nature. The sparse nature strengthens our independence assumption which is the case for OTFS as described in Chapter 1.

4.1.2 Factor Graph Description

With the above assumption of independence, we describe the factor graph formed by the system to describe the belief propagation algorithm. As per the definition, the factor graph is a bipartite graph with two sets of nodes, in this case the received signal elements being one set called the check nodes and transmit constellation vector elements being the other set called the variable nodes.

There are no connections between the nodes of same set. The interconnections between the two set of nodes is defined by channel matrix $\mathbf{H}$. The check node $\mathbf{y}(d)$ is connected to all the variable nodes $\mathbf{x}(c)$ where $c \in I_d$ and the variable node $\mathbf{x}(c)$ is connected to all check nodes $\mathbf{y}(d)$ where $d \in J_c$ as shown in Figure 4.1. The messages are passed from one set of nodes to other in each iteration which will be discussed below.

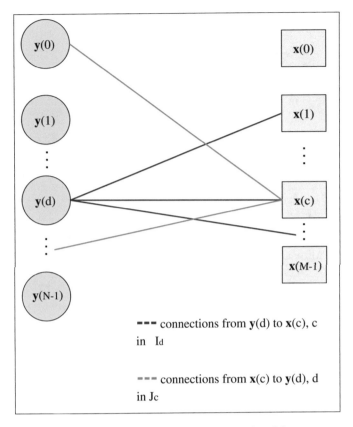

Figure 4.1 Factor graph for BP algorithm.

4.1.3 Equalization Algorithm

Message passing algorithm is an iterative algorithm in which messages are passed between check nodes and variable nodes in each iteration. It should be noted that the messages passed are the extrensic messages to the receiver node, i.e., the transmit node does not pass the information which the receiver node already knows. Here, the variable node passes the probability mass function (PMF) vector of itself to the check node while the check node passes the mean and variance of an assumed Gaussian PDF to the variable node which will be elaborated below. The algorithm has four steps namely, initiation, check node update, variable node update, and termination.

4.1.3.1 Initiation

In the intiation step, we define the initial message from variable node $\mathbf{x}(c)$ to $\mathbf{y}(d), d \in J_c$, $P_{c,d}^{(0)}$ as,

$$P_{c,d}^{(0)}(a_j) = \frac{1}{Q}, \ \forall a_j \in \mathbb{A}. \tag{4.5}$$

where, $Q = \|\mathbb{A}\|$ is the cardinality of constellation set. The above initiation because of the assumption that every constellation symbol is equally-likely .

4.1.3.2 Check Node Update

From (4.1),

$$\mathbf{y}(d) = \sum_{c \in I_d} \mathbf{H}(d,c)\mathbf{x}(c) + \mathbf{v}(d), \ d = 1, 2, \cdots N. \tag{4.6}$$

Since we are interested in finding $P(\mathbf{y}(d)|\mathbf{x}(c), \mathbf{H})$, the above expression can be simplified as,

$$\mathbf{y}(d) = \mathbf{H}(d,c)\mathbf{x}(c) + \sum_{e \in I_d, e \neq c} \mathbf{H}(d,e)\mathbf{x}(e) + \mathbf{v}(d)$$

$$\mathbf{y}(d) = \mathbf{H}(d,c)\mathbf{x}(c) + \zeta_{d,c}^i \tag{4.7}$$

where, $\zeta_{d,c}^i = \sum_{e \in I_d, e \neq c} \mathbf{H}(d,e)\mathbf{x}(e) + \mathbf{v}(d)$.

So, we can conclude $P(\mathbf{y}(d)|\mathbf{x}(c), \mathbf{H}) = p(\zeta_{d,c}^i)$. Since $\mathbf{v}(d) \sim N(0, \sigma_v^2)$, we apply central limit theorem and assume that $\zeta_{d,c}^i$ follow normal distribution, i.e., $\zeta_{d,c}^i \sim N(\mu_{d,c}^i, \sigma_{d,c}^{2(i)})$.

The mean can be calculated as,

$$\mu_{d,c}^i = E\{\zeta_{d,c}^i\}$$

$$\mu_{d,c}^i = E\{\sum_{e \in I_d, e \neq c} \mathbf{H}(d,c)\mathbf{x}(c) + \mathbf{v}(d)\}$$

$$\mu_{d,c}^i = E\{\sum_{e \in I_d, e \neq c} \mathbf{H}(d,c)\mathbf{x}(c)\} + E\{\mathbf{v}(d)\}$$

$$\mu_{d,c}^i = \sum_{e \in I_d, e \neq c} E\{\mathbf{H}(d,c)\mathbf{x}(c)\} \tag{4.8}$$

$$\mu_{d,c}^i = \sum_{e \in I_d, e \neq c} \mathbf{H}(d,c)E\{\mathbf{x}(c)\}$$

$$\mu_{d,c}^i = \sum_{e \in I_d, e \neq c} \mathbf{H}(d,c)\Big(\sum_{a_j \in A} P_{e,d}^{(i))}(a_j)a_j\Big)$$

And the variance can be given as,

$$\sigma_{d,c}^{2(i)} = E\{(\zeta_{d,c}^i)^2\} - (E\{\zeta_{d,c}^i\})^2$$

$$\sigma_{d,c}^{2(i)} = E\{(\sum_{e \in I_d, e \neq c} \mathbf{H}(d,c)\mathbf{x}(c) + \mathbf{v}(d))^2\} - (\mu_{d,c}^i)^2$$

$$\sigma_{d,c}^{2(i)} = E\{(\sum_{e \in I_d, e \neq c} \mathbf{H}(d,c)\mathbf{x}(c))^2 + 2(\sum_{e \in I_d, e \neq c} \mathbf{H}(d,c)\mathbf{x}(c))(\mathbf{v}(d)) + (\mathbf{v}(d))^2\} - (\mu_{d,c}^i)^2$$

$$\sigma_{d,c}^{2(i)} = E\{(\sum_{e \in I_d, e \neq c} \mathbf{H}(d,c)\mathbf{x}(c))^2\} + E\{(\mathbf{v}(d))^2\} - (\mu_{d,c}^i)^2$$

$$\sigma_{d,c}^{2(i)} = \sum_{e \in I_d, e \neq c} \|\mathbf{H}(d,c)\|^2 E\{\|\mathbf{x}(c)\|^2\} + \sigma_v^2 - (\mu_{d,c}^i)^2$$

$$\sigma_{d,c}^{2(i)} = \sum_{e \in I_d, e \neq c} \|\mathbf{H}(d,c)\|^2 \Big(\sum_{a_j \in A} P_{e,d}^{(i)}(a_j)\|a_j\|^2\Big) + \sigma_v^2 - \|\sum_{e \in I_d, e \neq c} \mathbf{H}(d,c)\Big(\sum_{a_j \in A} P_{e,d}^{(i)}(a_j)a_j\Big)\|^2$$

$$\tag{4.9}$$

Therefore, in iteration i, the pair of values $(\mu^i_{d,c}, \sigma^{2(i)}_{d,c})$ are passed as message from check node $\mathbf{y}(d)$ to variable node $\mathbf{x}(c)$).

4.1.3.3 Variable Node Update

At the variable node, the mean and variance messages are used to calculate the PMF vector for that variable node which is described below.

We know,

$$\tilde{P}_{c,d}^{\,i}(a_j) \propto \prod_{e \in J_c, e \neq d} \left(P(\mathbf{y}(d)|\mathbf{x}(e) = a_j, \mathbf{H}) \right), \; \forall a_j \in \mathbb{A} \tag{4.10}$$

The above expression holds true under independence of elements in received vector $\mathbf{y}$ as discussed earlier. Then,

$$\tilde{P}_{c,d}^{\,i}(a_j) \propto \prod_{e \in J_c, e \neq d} \left(\beta_{c,e,j} \right) \tag{4.11}$$

where $\beta_{c,e,j} = e^{\frac{-\|\mathbf{y}(e) - \mathbf{H}(e,c)\mathbf{a_j} - \mu^i_{\mathbf{e},\mathbf{c}}\|^2}{\sigma^{2(i)}_{e,c}}} \propto P(\mathbf{y}(d)|\mathbf{x}(e) = a_j, \mathbf{H})$

To obtain a valid PMF we define,

$$\tilde{P}_{c,d}^{\,i}(a_j) = \prod_{e \in J_c, e \neq d} \frac{\beta_{c,e,j}}{(\sum_j \beta_{c,e,j})} \tag{4.12}$$

Then we update the PMF vector using a weight factor Δ and memory as,

$$P_{c,d}^{(i)} = \Delta . \tilde{P}_{c,d} + (1 - \Delta) P_{e,d}^{(i-1)} \tag{4.13}$$

4.1.3.4 Criteria for Variable Node Decision Update

To determine whether to update the decision of transmitted symbols $\hat{\mathbf{x}}(c)$, we define convergence indicator $\eta^{(i)}$ as,

$$\eta^{(i)} = \frac{1}{N} \sum_{c=1}^{N} \mathbb{I}\left(\max_{a_j \in A}(p_c^{(i)}(a_j)) > (1 - \gamma) \right) \tag{4.14}$$

where, $\mathbb{I}(.)$ denotes indicator function, $\gamma > 0$ is threshold for checking convergence. Ideally γ should be small but very small values of γ can lead to nonconvergence. And the aggregate PMF element for variable node $\mathbf{x}(c)$, $p_c^i(a_j)$ is defined as,

$$p_c^i(a_j) = \prod_{e \in J_c} \frac{\beta_{c,e,j}}{(\sum_j \beta_{c,e,j})} \tag{4.15}$$

So, we update the decision if $\eta^{(i)} > \eta^{(i-1)}$, i.e., only when the current iteration gives better estimates than previous iteration given as,

$$\hat{\mathbf{x}}(c) = \underset{a_j \in \mathbb{A}}{\operatorname{argmax}} \, p_c^i(a_j) \tag{4.16}$$

4.1.3.5 Termination

The algorithm is instructed to stop if any of the following conditions is satified.

1. $\eta^{(i)} = 1$, i.e., at every variable node $\mathbf{x}(c)$, $p_c^i(a_j) = 1$ for some $a_j \in \mathbb{A}$. So, we can conclude from PMF that the $\hat{\mathbf{x}}(c) = a_j$ with probability 1.

2. $\eta(i) < \eta(j) - \epsilon$ for any $j \in \{1, 2, \cdots, i-1\}$, i.e., we terminate the algorithm in case the convergence indicator is decreased considerably compared to values in previous iterations. Choice of ϵ is important as it should not be very small considering there can be small fluctuations in convergence indicator values with itertion.

3. Number of iterations exceed maximum number of iterations N_{iter}.

4.1.4 Complexity Analysis

The number of operations required by the above algorithm for each iteration is given below.

Equation	Complexity
Means($\mu_{d,c}^i$) (4.8)	$O(NQS)$
Variance($\sigma_{d,c}^{2(i)}$) (4.9)	$O(NQS)$
PMF updation($P_{c,d}^i$) (4.13)	$O(MQS)$
Convergence indicator (η^i) (4.14)	$O(MQ)$
Decision/decoding ($\hat{\mathbf{x}}(c)$) (4.16)	$O(MQ)$

In the above table, $S = max(\{I_d|\forall d\}, \{J_c|\forall c\})$ is the maximum number of connections a node present in the factor graph. Hence, the complexity of this algorithm is $O(N_{iter} \max(M, N)QS)$.

4.1.5 Results

In Figure 4.2 and Figure 4.3, we present the BER performance of a RCP-OTFS system with reference results presented in [98] to evaluate the generalized implementation of the BP algorithm under similar evaluation environment. It can be observed that there is negligible difference between the implementation of the BP algorithm described above and the reference results. Thus, it can be concluded that the presented generalized BP is useful for sparse systems.

In Figure 4.4, the coded performance of RCP-OTFS system with the BP receiver presented in this chapter is compared with LMMSE receiver described in Section 4.2. It can be observed that the BP receiver outperforms LMMSE receiver by about 0.7 dB.

4.2 Low Complexity LMMSE Receiver for OTFS

In this section, a low complexity LMMSE receiver for OTFS, is described. The receiver exploits sparsity and quasi-banded structure of the matrices involved in the demodulation process of OTFS.

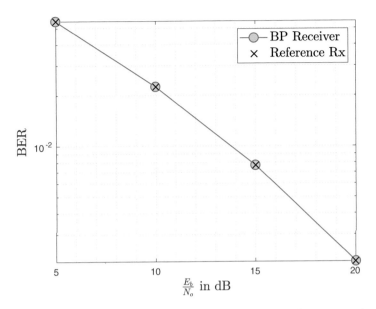

Figure 4.2 BER vs SNR for QPSK modulation for BP receiver.

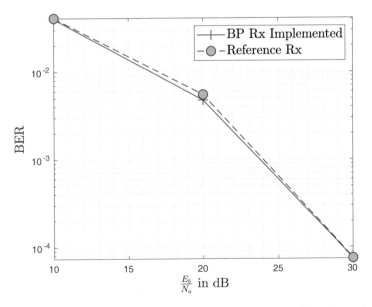

Figure 4.3 BER vs SNR for 16QAM modulation for BP receiver.

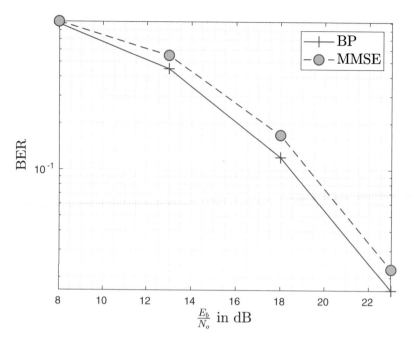

Figure 4.4   BLER vs SNR for 16QAM modulation.

4.2.1 Channel

The time varying channel for establishing the system equations for OTFS is described in Section 3.5. We also assume the perfect knowledge of (h_p, l_p, k_p), $p \in \mathbb{N}[0 \ P-1]$, at the receiver as in [99, 100, 101, 102, 103, 104]. Let τ_{max} and ν_{max} be the maximum delay and Doppler spread. Channel delay length $\alpha = \lceil \tau_{max} M \Delta f \rceil$ and channel Doppler length, $\beta = \lceil \nu_{max} NT \rceil$. Typically, $\alpha << MN$ as well as $\beta << MN$ which dictates the system matrices to be sparse (as will be seen in Sec. 4.2.2.1). For example, if we consider an OTFS system with $\Delta f = 15$ KHz, carrier frequency, $f_c = 4$ GHz, $N = 128$ and $M = 512$. We take a 3GPP vehicular channel EVA [30] with vehicular speed of 500 Kmph. De-Do channel lengths can be computed as, $\alpha = 20 << MN = 65536$ and $\beta = 16 << MN = 65536$.

4.2.2 Low Complexity LMMSE Receiver Design for OTFS

When $g(t)$ is rectangular, $\mathbf{A}$ in (3.6) is unitary. Thus, (3.23) is simplified to,

$$\hat{\mathbf{d}} = \mathbf{A}^\dagger \overbrace{\mathbf{H}^\dagger [\mathbf{H}\mathbf{H}^\dagger + \frac{\sigma_\nu^2}{\sigma_d^2}\mathbf{I}]^{-1}}^{\mathbf{H}_{eq}} \underbrace{\phantom{\mathbf{H}^\dagger [\mathbf{H}\mathbf{H}^\dagger + \frac{\sigma_\nu^2}{\sigma_d^2}\mathbf{I}]^{-1}}}_{\mathbf{r}_{ce}=\mathbf{H}_{eq}\mathbf{r}} \mathbf{r} \,. \tag{4.17}$$

Thus, LMMSE equalization can be performed as a two-stage equalizer. In the first stage, LMMSE channel equalization is performed to obtain $\mathbf{r}_{ce} = \mathbf{H}_{eq}\mathbf{r}$. The second stage

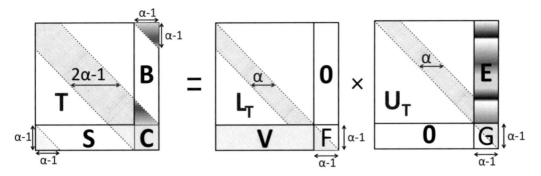

Figure 4.5 Structure of $\Psi = \mathbf{H}\mathbf{H}^\dagger + \frac{\sigma_\nu^2}{\sigma_d^2}\mathbf{I}$ matrix and its *LU* factorization.

is a OTFS matched filter receiver to obtain $\tilde{\mathbf{d}} = \mathbf{A}^\dagger \mathbf{r}_{ce}$. We will show in Sec. 4.2.2.3 that *the implementation of $\tilde{\mathbf{d}} = \mathbf{A}^\dagger \mathbf{r}_{ce}$ is simple which requires* $\frac{MN}{2}\log_2(N)$ *complex multiplications (CMs)*. But direct implementation of $\mathbf{r}_{ce} = \mathbf{H}_{eq}\mathbf{r}$ requires inversion of $\Psi = \mathbf{H}\mathbf{H}^\dagger + \frac{\sigma_\nu^2}{\sigma_d^2}\mathbf{I}$ and multiplication of $\mathbf{H}^\dagger$ which need $O(M^3N^3)$ CMs. Thus, we need to reduce the complexity of $\mathbf{r}_{ce} = \mathbf{H}_{eq}\mathbf{r}$. To do so, we investigate the structure of matrices involved in channel equalization below.

4.2.2.1 Structure of $\Psi = [\mathbf{H}\mathbf{H}^\dagger + \frac{\sigma_\nu^2}{\sigma_d^2}\mathbf{I}]$

Using (3.21), $\mathbf{H}\mathbf{H}^\dagger$ can be expressed as,

$$\mathbf{H}\mathbf{H}^\dagger = \sum_{p=1}^{P} h_p \Delta^{k_p} \Pi^{l_p} \sum_{s=1}^{P} \bar{h}_s \Delta^{-k_s} \Pi^{-l_s}. \tag{4.18}$$

Since Π is a circulant matrix, it can be verified that $\Pi^{l_p} = \mathbf{W}\Delta^{-l_p}\mathbf{W}^\dagger$. Therefore,

$$\mathbf{H}\mathbf{H}^\dagger = \sum_{\substack{p=1 \\ p=s}}^{P} |h_p|^2 \mathbf{I} + \sum_{\substack{p=1 \\ p\neq s}}^{P}\sum_{s=1}^{P} h_p \bar{h}_s e^{j\frac{2\pi}{MN}(l_p-l_s)} \Delta^{k_p-k_s} \Pi^{l_p-l_s} \tag{4.19}$$

Using (4.19), Ψ becomes,

$$\Psi = \sum_{\substack{p=1 \\ p=s}}^{P} (|h_p|^2 + \frac{\sigma_\nu^2}{\sigma_d^2})\mathbf{I} + \sum_{\substack{p=1 \\ p\neq s}}^{P}\sum_{s=1}^{P} h_p \bar{h}_s e^{j\frac{2\pi}{MN}(l_p-l_s)} \Delta^{k_p-k_s} \Pi^{l_p-l_s}. \tag{4.20}$$

Following (4.20), it can be concluded that the maximum shift of diagonal elements in Δ can be $\pm(\alpha - 1)$. Additionally due to the cyclic nature of shift, Ψ is quasi-banded with bandwidth of $2\alpha - 1$ as depicted in Figure 4.5. As discussed in Sec. 4.2.1, $\alpha << MN$, Ψ is also sparse for typical wireless channel. Since, we need to implement Ψ^{-1} in order to realize LMMSE receiver, we propose a low complexity LU decomposition of Ψ below.

Algorithm 1 Computation of $\mathbf{Y} = \mathbf{\Gamma}^{-1}\mathbf{X}$

1: Given : a lower triangular matrix $\mathbf{\Gamma}_{Q \times Q}$ and $\mathbf{X}_{Q \times \theta}$
2: Output : $\mathbf{Y}_{Q \times \theta} = \mathbf{\Gamma}_{Q \times Q}^{-1}\mathbf{X}_{Q \times \theta}$
3: **for** $s = 0 : \theta$ **do**
4:      $\mathbf{Y}(0, s) = \frac{\mathbf{X}(0,s)}{\mathbf{\Gamma}(0,0)}$
5:      **for** $k = 1 : \theta$ **do**
6:          $\mathbf{Y}(k, s) = \frac{1}{\mathbf{\Gamma}(k,k)}\mathbf{X}(k, s) - \sum_{i=1}^{k-1} \mathbf{\Gamma}(k, k - i)\mathbf{Y}(k - i, s)$
7:      **end for**
8:      **for** $k = \theta + 1 : Q$ **do**
9:          $\mathbf{Y}(k, s) = \frac{1}{\mathbf{\Gamma}(k,k)}\mathbf{X}(k, s) - \sum_{i=1}^{P-1} \mathbf{\Gamma}(k, k - i)\mathbf{Y}(k - i, s)$
10:     **end for**
11: **end for**

4.2.2.2 Low Complexity LU Factorization of Ψ

To implement the low complexity LU factorization of Ψ, we propose following partition of Ψ (by considering, $\theta = \alpha - 1$ and $Q = MN - \theta$).

$$\underbrace{\begin{bmatrix} \mathscr{T}_{Q \times Q} & \mathbf{B}_{Q \times \theta} \\ \mathbf{S}_{\theta \times Q} & \mathbf{C}_{\theta \times \theta} \end{bmatrix}}_{\Psi} = \underbrace{\begin{bmatrix} \mathscr{L}_{Q \times Q} & \mathbf{0}_{Q \times \theta} \\ \mathbf{V}_{\theta \times Q} & \mathbf{F}_{\theta \times \theta} \end{bmatrix}}_{\mathbf{L}} \times \underbrace{\begin{bmatrix} \mathscr{U}_{Q \times Q} & \mathbf{E}_{Q \times \theta} \\ \mathbf{0}_{\theta \times Q} & \mathbf{G}_{\theta \times \theta} \end{bmatrix}}_{\mathbf{U}} \tag{4.21}$$

Using the partition in (4.21), following equalities hold

$$\mathscr{T} = \mathscr{L}\,\mathscr{U} \tag{4.22}$$

$$\mathbf{E} = \mathscr{L}^{-1}\,\mathbf{B} \tag{4.23}$$

$$\mathbf{V} = \mathbf{S}\,\mathscr{U}^{-1} \tag{4.24}$$

$$\mathbf{FG} = \mathbf{C} - \mathbf{VE} \tag{4.25}$$

Next, we discuss the solution of (4.22-4.25) to compute LU factorization of Ψ. Since, $\mathscr{T}$ is a banded matrix its LU decomposition can be computed using low complexity algorithm presented in [105]. $\mathscr{L}^{-1}\mathbf{B}$ can be computed using forward substitution algorithm for lower triangular banded matrix as explained in Algorithm 1. One can compute (4.24) in following two steps. As, $\mathscr{U}^\dagger$ is a lower triangular banded matrix, in the first step, we compute $\mathbf{V}^\dagger = (\mathscr{U}^\dagger)^{-1}\mathbf{S}^\dagger$ using Algorithm 1. Finally, $\mathbf{V}$ can be computed simply by taking hermitian of $\mathbf{V}^\dagger$. As $\theta << MN$, even a direct computation of (4.25) requires $O(\theta^2 MN)$ computations. As, $\mathbf{F}$ is a lower triangular matrix and $\mathbf{G}$ is an upper triangular matrix, $\mathbf{F}$ and $\mathbf{G}$ can be computed using LU decomposition of (4.25). Pivotal Gaussian elimination algorithm [106] can be used to compute LU decomposition of (4.25) without much increase in complexity. It should be noted that diagonal values of $\mathscr{L}$ and $\mathbf{F}$ are unity. Thus, diagonal values of $\mathbf{L}$ are also unity.

Note on the nonsingularity of L and U

For LMMSE processing, $\mathbf{L}$ and $\mathbf{U}$ need to be inverted (as will be discussed in Sec. 4.2.2.3). We next discuss the non-singularity of $\mathbf{L}$ and $\mathbf{U}$. As $\mathbf{HH}^\dagger$ is a hermitian matrix, its a positive

---

Algorithm 2 Computation of $\mathbf{r}^{(1)} = \mathbf{L}^{-1}\mathbf{r}$

1: Given : a quasi-banded lower triangular matrix $\mathbf{L}_{MN \times MN}$ and $\mathbf{r}_{MN \times 1}$
2: Output : $\mathbf{r}^{(1)}_{MN \times 1} = \mathbf{L}^{-1}_{MN \times MN}\mathbf{r}_{MN \times 1}$
3: $\mathbf{r}^{(1)}(0) = \mathbf{r}(0)$
4: **for** $k = 1 : \alpha - 1$ **do**
5: $\quad \mathbf{r}^{(1)}(k) = \mathbf{r}(k) - \sum_{i=1}^{k-1} \mathbf{L}(k, k-i)\mathbf{r}^{(1)}(k-i)$
6: **end for**
7: **for** $k = \alpha : Q$ **do**
8: $\quad \mathbf{r}^{(1)}(k) = \mathbf{r}(k) - \sum_{i=1}^{\alpha-1} \mathbf{L}(k, k-i)\mathbf{r}^{(1)}(k-i)$
9: **end for**
10: **for** $k = Q + 1 : MN - 1$ **do**
11: $\quad \mathbf{r}^{(1)}(k) = \mathbf{r}(k) - \sum_{i=1}^{MN-1} \mathbf{L}(k, k-i)\mathbf{r}^{(1)}(k-i)$
12: **end for**

Algorithm 3 Computation of $\mathbf{r}^{(2)} = \mathbf{U}^{-1}\mathbf{r}^{(1)}$

1: Given : a quasi-banded upper triangular matrix $\mathbf{U}_{MN \times MN}$ and $\mathbf{r}^{(1)}_{MN \times 1}$
2: Output : $\mathbf{r}^{(2)}_{MN \times 1} = \mathbf{U}^{-1}_{MN \times MN}\mathbf{r}^{(1)}_{MN \times 1}$
3: $\mathbf{r}^{(2)}(MN - 1) = \frac{\mathbf{r}^{(1)}(MN-1)}{\mathbf{U}(MN-1, MN-1)}$
4: **for** $k = MN - 2 : MN - 2\alpha$ **do**
5: $\quad \mathbf{r}^{(2)}(k) = \frac{1}{\mathbf{U}(k,k)}\mathbf{r}^{(1)}(k) - \sum_{i=1}^{MN-k-1} \mathbf{U}(k, k+i)\mathbf{r}^{(2)}(k+i)$
6: **end for**
7: **for** $k = \alpha : Q$ **do**
8: $\quad \mathbf{r}^{(2)}(k) = \frac{1}{\mathbf{U}(k,k)}\mathbf{r}^{(1)}(k) - \sum_{i=1}^{\alpha} \mathbf{U}(k, k+i)\mathbf{r}^{(2)}(k+i) - \sum_{r=MN-\alpha}^{MN-1} \mathbf{U}(k, r)\mathbf{r}^{(2)}(r)$
9: **end for**

---

semidefinite matrix. Since $\frac{\sigma_x^2}{\sigma_d^2} > 0$ for finite SNR ranges, Ψ is a positive definite matrix; therefore, Ψ is invertible. As diagonal values of $\mathbf{L}$ are unity, $\mathbf{L}$ is nonsingular. Further, nonsingularity of $\mathbf{U}$ is a consequence of nonsingularity of Ψ [106].

4.2.2.3 Computation of $\hat{\mathbf{d}}$

After LU decomposition of Ψ, $\mathbf{r}_{ce}$ is simplified to,

$$\mathbf{r}_{ce} = \mathbf{H}^\dagger \overbrace{\mathbf{U}^{-1} \underbrace{\mathbf{L}^{-1}\mathbf{r}}_{\mathbf{r}^{(1)}}}^{\mathbf{r}^{(2)}}. \tag{4.26}$$

As $\mathbf{L}$ is a quasi-banded lower triangular matrix, $\mathbf{r}^{(1)} = \mathbf{L}^{-1}\mathbf{r}$ can be computed using low complexity forward substitution as explained in Algorithm 2. $\mathbf{r}^{(2)} = \mathbf{U}^{-1}\mathbf{r}^{(1)}$ can be computed using Algorithm 3.

Using the definition of $\mathbf{H}$, $\mathbf{r}_{ce} = \mathbf{H}^\dagger\mathbf{r}^{(2)}$ can be given as,

$$\mathbf{r}_{ce} = \sum_{p=1}^{P} \bar{h}_p \Delta^{-k_p} \underbrace{\Pi^{-l_p}\mathbf{r}^{(2)}}_{\text{circular shift}} \tag{4.27}$$

To compute $\mathbf{r}_{ce}$, $\mathbf{r}^{(2)}$ is first circularly shifted by delay $-l_p$ and then multiplied by $\bar{h}_p \text{diag}\{\Delta^{-k_p}\}$ by using point-to-point multiplication for each path p. All vectors obtained in above step are finally summed to obtain $\mathbf{r}_{ce}$.

Instead of directly computing $\hat{\mathbf{d}}$ as $\mathbf{A}^\dagger \mathbf{r}_{ce}$, we first reshape $\mathbf{r}_{ce}$ to a $M \times N$ size $\mathbf{R}$ matrix as,

$$\mathbf{R} = \begin{bmatrix} \mathbf{r}_{ce}(0) & \mathbf{r}_{ce}(M) & \cdots & \mathbf{r}_{ce}(MN-N) \\ \mathbf{r}_{ce}(1) & \mathbf{r}_{ce}(M+1) & \cdots & \mathbf{r}_{ce}(MN-N+1) \\ \vdots & \vdots & \ddots & \vdots \\ \mathbf{r}_{ce}(M-1) & \mathbf{r}_{ce}(2M-1) & \cdots & \mathbf{r}_{ce}(MN-1) \end{bmatrix}. \tag{4.28}$$

Then, we perform

$$\hat{\mathbf{d}} = vec\{\mathbf{R}\mathbf{W}_N^\dagger\}, \tag{4.29}$$

which can be implemented using M number of N-point FFT operations. Figure 4.6 describes the signal processing steps of our proposed low complexity LMMSE receiver.

4.2.2.4 LMMSE Receiver for OFDM over TVC

Low complexity receiver discussed for OTFS can easily be extended to OFDM by setting $\mathbf{A} = \mathbf{I}_N \otimes \mathbf{W}_M$. To do so, $\mathbf{r}_{ce} = \mathbf{H}_{eq}\mathbf{r}$ is performed by computing (4.21-4.27) as discussed in Sec. 4.2.2.2 and 4.2.2.3. Further, $\hat{\mathbf{d}} = (\mathbf{I}_N \otimes \mathbf{W}_M^\dagger)\mathbf{r}_{ce}$ can be computed using N number of M-point FFTs.

Justification on considering rectangular pulse shaped OTFS

In a general OTFS setting, transmitter and receiver pulses can be different. We consider g_{tx} and g_{rx} be transmitter and receiver pulses, respectively. These pulses need not necessarily be orthogonal. Suppose OTFS is transmitted through an LTV channel having maximum delay spread and maximum Doppler spread of τ_{max} and ν_{max}, respectively. Considering matched filter receiver, these pulses are called ideal if they satisfy biorthogonality property for $(n, m)^{\text{th}}$ data symbol, $n \in \mathbb{N}[0 \ M-1]$ and $m \in \mathbb{N}[0 \ N-1]$, [101], i.e.,

$$\mathscr{A}_{g_{rx},g_{tx}}(t,f)|_{t=nT+(-\tau_{max} \ \tau_{max}), \ f=m\Delta f+(-\nu_{max} \ \nu_{max})}$$
$$= \delta(n)\delta(m)U[-\tau_{max} \ \tau_{max}]U[-\nu_{max} \ \nu_{max}], \tag{4.30}$$

where, $\mathscr{A}_{g_{rx},g_{tx}}(t,f) = \int_{\mathbb{R}} g_{rx}(t+\frac{\tau}{2})\overline{g_{tx}(t-\frac{\tau}{2})}e^{-j2\pi t f}dt$ is cross ambiguity function and

$$U[a \ b](x) = \begin{cases} 1 \text{ If } a < x < b \\ 0 \text{ otherwise.} \end{cases}$$. To satisfy biorthogonality property, cross ambiguty

function should be confined in a rectangular area of $4\tau_{max}\nu_{max}$ which is impossible for a general wireless channel because of,

1. considerations of under-spread wireless channel for which $4\tau_{max}\nu_{max} < 1$ [23],
2. and TF uncertainty principle [29] which states that the minimum area covered by a pulse in TF domain is one which is satisfied by Gaussian pulse..

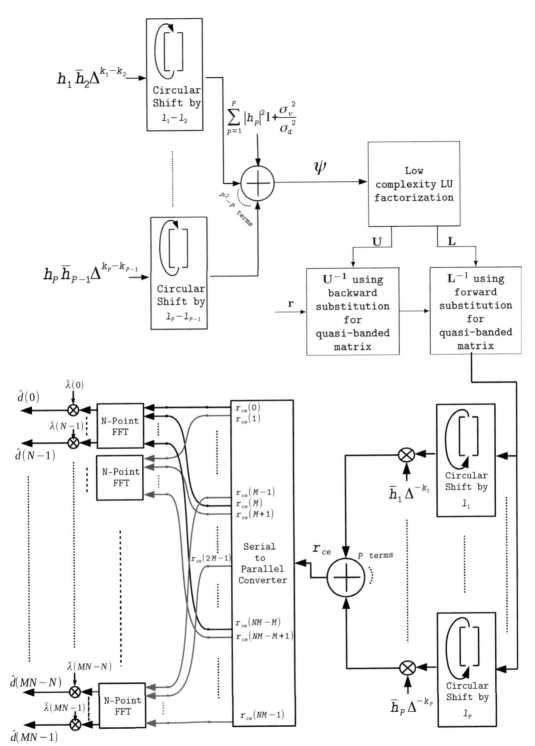

Figure 4.6 Our proposed low complexity OTFS-MMSE receiver.

Table 4.1 Computational complexity of different operations in our proposed receiver.

Operation	Number of complex multiplications
(4.20)	$[P^2 - P][2\beta + 1]MN + P$
(4.22)	$[\alpha^2 + 2\alpha]MN$
(4.23,4.24) using Algorithm 1	$\alpha MN - \frac{3\alpha^3 + \alpha}{2}$
(4.25) and LU decomposition of **FG**	$\alpha^2 MN - MN + \frac{2\alpha^3}{3}$
Algorithm 2 and 3	$MN[2\alpha - 1] + \frac{3\alpha^2}{2} + \frac{\alpha}{2}$
(4.27)	$P(\beta + 1)MN$
(4.29)	$\frac{MN}{2} \log_2(N)$

Table 4.2 Computational complexity of different receivers.

Structure	Number of complex multiplications
OFDM receiver direct using (3.23)	$\frac{MN}{2} \log_2 M + \frac{8}{6}(MN)^3 + 2(MN)^2$
OTFS receiver direct using (3.23)	$\frac{MN}{2} \log_2 N + \frac{8}{6}(MN)^3 + 2(MN)^2$
Described OFDM receiver	$\frac{MN}{2} \log_2 M + MN[2\alpha^2 + 2P^2\beta + 9\alpha - P\beta - 3] + \frac{2}{3}\alpha^3 + 2\alpha + P$
Described OTFS receiver	$\frac{MN}{2} \log_2 N + MN[2\alpha^2 + 2P^2\beta + 9\alpha - P\beta - 3] + \frac{2}{3}\alpha^3 + 2\alpha + P$

Thus, it can be concluded that ideal pulse does not exist. For example, if we consider an OTFS system with $\Delta f = 15$ KHz, carrier frequency, $f_c = 4$ GHz, $N = 128$ and $M = 512$. We take a 3GPP vehicular channel EVA [30] with vehicular speed of 500 Kmph. For this system $\tau_{max} = 2.5\mu s$, $\nu_{max} = 2.7$ KHz and thus $4\tau_{max}\nu_{max} = 0.027 << 1$.

In a more general way, our proposed receiver is applicable to any orthogonal waveform. Rectangular pulse shape which is generally considered in the OTFS literature for practical implementation is also orthogonal. It should be noted that our proposed receiver is not applicable for nonorthogonal waveform i.e., when $\mathbf{AA}^\dagger \neq \mathbf{I}$.

4.2.3 Result

4.2.3.1 Computational Complexity

In this section, we present the computational complexity of the low complexity LMMSE receiver. We calculate the complexity in terms of total number of complex multiplications (CMs). c-point FFT and IFFT can be implemented using radix-2 FFT algorithm using $\frac{c}{2} \log_2(c)$ CMs [85]. The complexity of the proposed receiver can be computed using the structure provided in Sec. 4.2.2.2. Computation of $c \times c$ matrix-matrix multiplication, matrix inversion, and LU decomposition require $\frac{c^3}{2}$, $\frac{2c^3}{3}$ and $\frac{2c^3}{3}$ CMs, respectively. Total CMs required to compute different operations in the low complexity receiver are presented in Table 4.1. CMs required for different receivers is presented in Table 4.2. It is evident that the low complexity LMMSE receiver has complexity of $O(MN[\log_2(N) + \alpha^2 + P^2\beta])$.

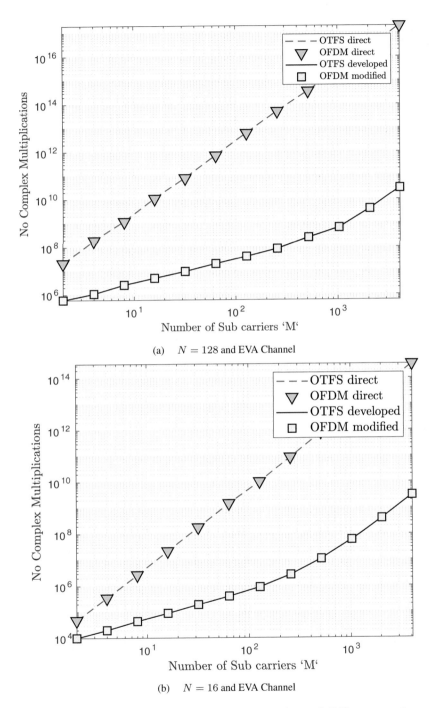

(a) $N = 128$ and EVA Channel

(b) $N = 16$ and EVA Channel

Figure 4.7   Computation complexity comparison of different receivers.

Table 4.3   Simulation parameters.

Number of subcarriers M	512
Number of time slots N	128
Mapping	4 QAM
Subcarrier bandwidth	15 KHz
Channel	EVA [30]
Vehicular speed	500 Kmph
Carrier frequency	4 GHz

To evaluate the complexity reduction achieved by our proposed receiver, we consider an OTFS system with $\Delta f = 15\ KHz$, $f_c = 4\ GHz$ and vehicular speed of 500 Kmph. We consider two 3GPP vehicular channel models [30] namely: (i) extended vehicular A (EVA) with $P = 9$ and $\tau_{max} = 2.51\ \mu\ sec$, and (ii) extended vehicular B (EVB) with $P = 6$ and $\tau_{max} = 20\ \mu\ sec$. Two block durations are assumed namely (i) small block with $N = 16$ and $T_f = 1.1\ msec$. and (ii) large block with $N = 128$ and $T_f = 8.85\ msec$. The complexity presented in Table 4.2 is plotted in figures4.7a and 4.7b for $M \in [2\ 4096]$. It is evident from the figure that for EVA channel the low complexity receivers require upto 10^7 and 10^5 times lower CMs than direct ones using (3.23) for large and small block, respectively. Whereas, for EVB channel, (not shown here) the low complexity receiver needs 2.5×10^5 and 3000 times lesser CMs over the direct ones using (3.23) for large and small block, respectively. This reduction in complexity gain for EVB channel as compared with EVA channel is due to increase in α. We can conclude that our proposed receivers achieve a significant complexity reduction over direct implementation of (3.23).

4.2.3.2 BER Evaluation

Here, we present BER performance of the low complexity receiver in EVA channel. Simulation parameters are given in Table 4.3. Doppler is generated using Jake's formula, $\nu_p = \nu_{max} cos(\theta_p)$, where θ_p is uniformly distributed over $[-\pi\ \pi]$. The CP is chosen long enough to accommodate the wireless channel delay spread. We plot BER for different waveforms in Figure 4.8 and 4.9 for 500 Kmph and 30 Kmph vehicular speed, respectively. Direct implementation of the receivers are obtained by using (3.23). It can be observed that the low complexity receivers do not incur performance loss when compared with direct implementation of receivers. It can also be observed that OTFS-LMMSE receiver can extract diversity gain, for instance at the BER of 5×10^{-4}, OTFS-LMMSE receiver achieves an SNR gain of 13 dB over OFDM-MMSE receiver.

4.3 Iterative Successive Interference Cancellation Receiver

4.3.1 Introduction

In this part of the book, we focus on RCP-OTFS as it has higher spectral efficiency in comparison with CP-OTFS. The view that RCP-OTFS can also be seen as an interleaved OFDM system is established in Chapter 3. It is also true that most of the known broadband

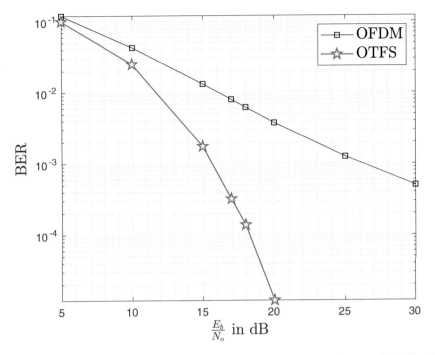

Figure 4.8 The BER performance of different waveforms for $v = 500$ Kmph.

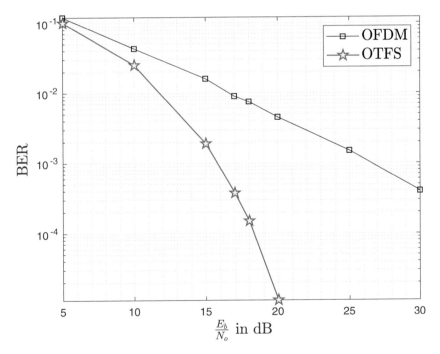

Figure 4.9 The BER performance of different waveforms for $v = 30$ Kmph.

wireless communication systems use forward error correcting codes (FEC) in order to achieve desirable performance over wireless channel [107].

At the transmitter low density parity check (LDPC) code based block channel encoder is considered to be available. Block FECs have inherent capability to detect incorrect code blocks. This property of block FEC is exploited to design the iterative successive interference cancellation (SIC) based receiver. The received signal is first equalized by linear minimum mean square error (LMMSE) equalizer (Section 4.2). The soft outputs of which are fed to the LDPC decoder. Such decoders not only provide corrected code blocks but also indicate the code blocks which are not correctly decodable. The correct code blocks are used to regenerate interference patterns, which is cancelled from the received signal. This signal which are less contaminated with interference is used for the next iteration. Since, OTFS suffers from strong interference when it undergoes a time varying channel, the above procedure is iterated through in order to minimize the effect of such interference.

4.3.2 LDPC Coded LMMSE-SIC Reciever

The over signal flow the SIC receiver based on LMMSE structure is shown in figure 4.10. We consider $\mathbf{r}_i^{int}$ contain interference pattern for i^{th} iteration, where $1 \leq i \leq N_{SIC}$. For the first iteration of SIC receiver, $\mathbf{r}_1^{int} = 0$. Interference free signal for i^{th} iteration is generated using,

$$\mathbf{r}_i = \mathbf{r} - \mathbf{r}_i^{int}. \tag{4.31}$$

Next, $\mathbf{r}_i$ is equalized using LMMSE equalizer as,

$$\hat{\mathbf{d}}^{(i)} = \mathbf{A}^\dagger \mathbf{H}^\dagger [\mathbf{H}\mathbf{H}^\dagger + \tfrac{\sigma_\nu^2}{\sigma_d^2}\mathbf{I}]^{-1}\mathbf{r}_i. \tag{4.32}$$

These soft values of $\hat{\mathbf{d}}^{(i)}$ are used by LDPC [108] decoder to provide estimated values of bits $\hat{b}_i$. We let the indices of correct blocks after i^{th} iteration be stored in the vector $\mathbf{indexc}^{(i)}$ with $\mathbf{indexc}^{(1)} = \mathbf{0}$. To avoid unnecessary computations, we perform LDPC decoding only for those code blocks which were incorrectly decoded in the previous iteration. To do so, we set $\hat{\mathbf{d}}^{(i)}[\mathbf{indexinc}] = \tilde{\mathbf{d}}^i$. To decode the equalized data corresponding to only incorrect blocks in the last iteration using LDPC decoder, the log-likelihood ratios(LLRs) of $\tilde{\mathbf{d}}^i$ are passed to the decoder which are calculated from the equalized symbols as,

$$LLR(b_\eta^j|\hat{x}(\eta)) \approx (\min_{s\epsilon S_j^0}\frac{||\hat{x}(\eta) - s||^2}{\sigma^2(\eta, \eta)}) - (\min_{s\epsilon S_j^1}\frac{||\hat{x}(\eta) - s||^2}{\sigma^2(\eta, \eta)}) \tag{4.33}$$

where $\hat{x}(\eta)$ is the ηth element of $\hat{x}$ mapped from the bits $b_\eta^0 b_\eta^1 \cdots b_\eta^{J-1}$, J is the number of bits per symbol and $\sigma^2(\eta, \eta)$ is the element of $\sigma^2 = \sigma_n^2(\mathbf{H}_{\mathbf{mmse}}\mathbf{H}_{\mathbf{mmse}}^\dagger)$, where $\mathbf{H}_{mmse} = \mathbf{A}^\dagger\mathbf{H}^\dagger[\mathbf{H}\mathbf{H}^\dagger + \tfrac{\sigma_\nu^2}{\sigma_d^2}\mathbf{I}]^{-1}$. S_j^0 and S_j^1 denote the set of constellation symbols, where the bit $b_\eta^j = 0$ and $b_\eta^j = 1$, respectively for $j = 0, 1, \cdots, J - 1$.

These LLRs are then fed into the LDPC decoder to decode data. Let $\mathbf{L}$ denotes a matrix where $\mathbf{L}(\eta, j) = LLR(b_\eta^j|\hat{x}(\eta))$ for $\eta = 1, 2, \cdots, MN$ and $j = 0, 1, \cdots, J - 1$. $\mathbf{L}$ is reshaped to $L_{cl} \times K$ matrix Kmph where L_{cl} and K denote the LDPC codeword length and number of codewords, respectively. Each column of **L** subsequently regenerates message word m_ι for $\iota = 1, 2 \cdots, K$ using the Min–Sum algorithm[109] employed by the LDPC decoder and is collected as the recovered data.

Let us say that the total number of blocks be K, then symbols per block $S = \frac{MN}{K}$, which we assume to be integers. Let us consider that the total number of incorrect blocks after i^{th} iteration is K_I^i. The index of incorrect blocks after i^{th} iterations are stored in the vector **indexinc**$^{(i)}$.

To increase the detectability of incorrect blocks in the next iteration, we intend to generate interference for incorrect blocks emanating from correct blocks. To do so, firstly, we regenerate QAM symbols from detected bits which is denoted by $\tilde{\mathbf{d}}_i$. Next, symbol values corresponding to incorrect symbols are set to zero i.e.,

$$\tilde{\mathbf{d}}_i[\mathbf{indexinc}^{(i)}] = \mathbf{0}. \tag{4.34}$$

To generate the interference pattern $\mathbf{r}_{i+1}^{int}$, $\tilde{\mathbf{d}}_i$ is modulated using OTFS modulation and multiplied by **H** i.e.,

$$\mathbf{r}_{i+1}^{int} = \mathbf{HA}\tilde{\mathbf{d}}_i. \tag{4.35}$$

Stopping criteria

Maximum number of iteration for our proposed receiver is N_{SIC}. We propose two additional stopping criteria. Suppose number of blocks in error after i^{th} iteration is N_e^i. Iterations will stop if there is an insignificant improvement in correcting the blocks between iterations. i.e., iterations will stop if $\text{imp}_{\text{fac}} = N_e^i - N_e^{i-1} \le \eta_1$, where η_1 is improvement tolerance constant. Another stopping criteria is to stop the loop when the number of blocks in error after a iteration reaches a very low value i.e., iterations will stop if $\text{err}_{\text{fac}} = N_e^i \le \eta_2$, where η_2 is error threshold value.

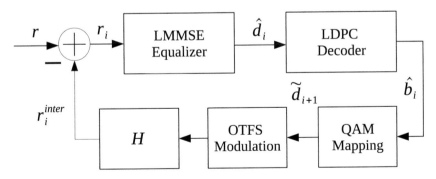

Figure 4.10 Schematic block diagram representing signal flow of our proposed code word level SIC receiver.

Algorithm 4 The developed SIC receiver

1:    Given : $\mathbf{r}$, $\mathbf{H}$, η_1 and η_2

2:    Output : $\hat{b}$

3:    Initialize : $imp_{fac} = 10\eta_1$ $err_{fac} = 10\eta_2$, $\mathbf{indexc^1} = \mathbf{0}$

4:    $\mathbf{r}_1^{int} = \mathbf{0}$ and $\tilde{\mathbf{d}}_1 = \mathbf{0}$

5:    Compute $\Psi = [\mathbf{HH}^\dagger + \frac{\sigma_\nu^2}{\sigma_d^2}\mathbf{I}]$

6:    Compute $\mathbf{LU} = \Psi$

7:    **while** ($i \leq N_{SIC}$ && $\mathrm{imp}_{\mathrm{fac}} \geq \eta_1$ && $\mathrm{err}_{\mathrm{fac}} \geq \eta_2$) **do**

8:      $\mathbf{r}_i = \mathbf{r} - \mathbf{r}_i^{int}$

9:      $\hat{\mathbf{d}}^{(i)} = \mathbf{A}^\dagger \mathbf{H}^\dagger \mathbf{U}^{-1} \mathbf{L}^{-1} \mathbf{r}_i$

10:     $\hat{\mathbf{d}}^{(i)}[\mathbf{indexinc}^{(i)}] = \tilde{\mathbf{d}}^i$

11:     $[\hat{\mathbf{b}}_i, \mathbf{indexinc}^{(i)}] = LDPCdecoder\,(\tilde{\hat{\mathbf{d}}}^{(i)})$

12:     $\mathbf{d}_{i+1} = QAMmod\,(\hat{\mathbf{b}}_i)$

13:     $\tilde{\mathbf{d}}_{i+1}[\mathbf{index_inc}^{(i)}] = \mathbf{0}$

14:     $\mathbf{r}_{i+1}^{int} = \mathbf{HA}\tilde{\mathbf{d}}_{i+1}$

15:     $\mathrm{imp}_{\mathrm{fac}} = N_e^i - N_e^{i-1}$

16:     $\mathrm{err}_{\mathrm{fac}} = N_e^i$

17: **end while**

18: $\hat{b} = \hat{b}_{N_{SIC}}$

Working principle of the iterative SIC receiver

Using (4.35), (4.31) can be written as,

$$\mathbf{r}_{i+1} = \mathbf{HA}(\mathbf{d} - \mathbf{d}_i) + \mathbf{n} = \mathbf{HA}\tilde{\mathbf{d}}_{i+1}. \tag{4.36}$$

It can be observed that $\tilde{\mathbf{d}}_{i+1}$ holds only data points corresponding to incorrect blocks in the previous iteration. This means that the interference due to correct blocks in i^{th} step is completely eliminated in $\mathbf{r}_{i+1}$. Thus, detectability of symbols increase with the iteration.

4.3.3 Low Complexity Receiver

The LMMSE equalization can be implemented using steps in section 4.2. The LMMSE implementation computes LU decomposition of $\mathbf{H}^\dagger[\mathbf{HH}^\dagger + \frac{\sigma_\nu^2}{\sigma_d^2}\mathbf{I}]$ which is independent of $\mathbf{r}_i$. Thus, LU decomposition needs to be computed only once and can be stored for further iterations of SIC receiver. Computation of (4.33) requires computation in the $O(MN)$ [109]. Further, (4.35) can be implemented using M number of IFFTs [99] and αMN complex multiplications[94].

4.3.3.1 Complexity Computation

In this section, we present the computational complexity of the iterative SIC receiver. To do so, we compute the total number of complex multiplications needed to implement our proposed receiver. Table 4.4 presents the computational complexity of different receivers. Figure 4.11 presents cumulative distribution function (CDF) of SIC iterations vs SNR for 64 QAM modulation, $N = 128$, $M = 512$ and EVB channel. It can be observed that the number of iterations are decreasing with the increasing value of SNR. One may observe

Table 4.4   Computational complexity of different receivers.

Structure	Number of complex multiplications
OTFS LMMSE receiver [94] C_{LMMSE}	$\frac{MN}{2}\log_2 N + MN[2\alpha^2 + 2P^2\beta + 9\alpha - P\beta - 3] + \frac{2}{3}\alpha^3 + 2\alpha + P$
OTFS MP based receiver [101]	$N_{MP}MN(11QS + 5S + 2Q + 2 + 0.5log_2(N))$
Described SIC based receiver	$C_{LMMSE} + (N_{SIC} - 1)(0.5MN + P(\beta + 1)MN + 4\alpha MN - 2MN + 3\alpha^2 + \alpha)$

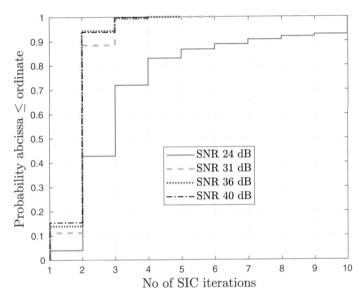

Figure 4.11   Cumulative distribution of SIC iterations for 64 QAM, $N = 128, M = 512$.

shifting of CDF curve towards left with increasing SNR. At higher SNR lesser number of code blocks are in error in the first stage itself, hence, such observations.

Computational complexities presented in Table 4.4 are plotted in Figure 4.12 for $M = 512$ and $N \in \mathbb{Z}[2\ 512]$. We compute worst case complexity for the iterative SIC receiver by considering that the receiver uses $N_{SIC} = 10$ iterations. In actual implementation, the number of iterations will be much less as shown in Figure 4.11. It can be observed that the SIC receiver has similar complexity to that of the MMSE receiver. This is mainly due to the one time LU decomposition. It can also be observed that the MP receiver in [101] is around ten and forty times more complex than our proposed receiver for 16-QAM and 64-QAM modulations, respectively. It may be noted that the complexity of SIC and LMMSE receivers are not affected by the modulation order in the transmitted signal. Thus, it is interesting to observe that despite being a nonlinear iterative receiver, its complexity is comparable to that of a linear receiver.

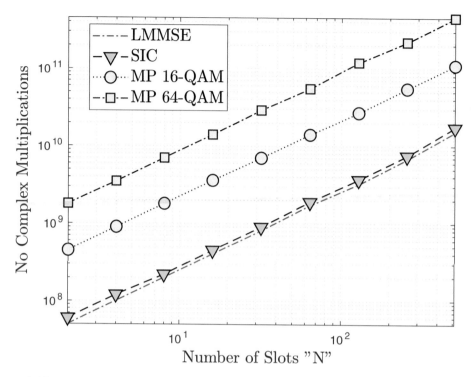

Figure 4.12   Computational complexity of OTFS receivers for $M = 512$ and EVB channel profile and vehicular speed of 500 Kmph. $N_{SIC} = 10$, number of iteration for MP receiver is 20.

4.3.4 Performance Presents Cumulative Distribution

The performance of the LMMSE-SIC receiver is discussed in this section. The parameters used for generating Monte–Carlo simulations are described in Table 4.5. Two different frame sizes are chosen, namely the large frame size , i.e., $N = 128$, $M = 512$ and the small frame size i.e $N = 16$, $M = 128$,

Uncoded bit error rate (BER) vs SNR of known receivers for OTFS is shown in Figure 4.13.

Let us consider $4 - QAM$ in a large frame. The performance of message passing (MP) algorithm, is obtained using the codes released by [101], however, with maximum allowed iteration of 50 and convergence factor of 0.6. It can be observed that it has better performance than LMMSE receiver by ~ 3 dB. For a small frame with $4 - QAM$, it is observed that both receivers degrade by ~ 3 dB.

From the performance curve of the above mentioned MP algorithm for the small frame with $16 - QAM$, one can observe that the appearance of error floor at higher SNRs. On the other hand the MMSE receiver for the same configuration is found to be ~ 2.5 dB poorer at a BER of 10^{-3}, but it is found to be superior at higher SNR as it does not encounter the error floor.

Table 4.5 Simulation parameters.

Number of subcarriers M	128 for short frame size and 512 for large frame size
Number of time slots N	16 for short frame size and 128 for large frame size
Mapping	4, 16 and 64 QAM
Subcarrier bandwidth	15 KHz
Channel	Extended vehicular channel -A (EVA) ($P = 9$ and $\tau_{max} = 2.51\mu s$) and EVB ($P = 6$ and $\tau_{max} = 20\mu s$) [30]
Vehicular speed	500 Kmph
Carrier frequency	4 GHz
FEC (LDPC)	Coderate 2/3, codeword length 648 [108]

The large frame with 16-QAM is found to provide a $\sim$ 2 dB gain for the MMSE algorithm compared to the small frame for the same algorithm.

Thus, it can be concluded from the uncoded BER curves that the MP algorithm has better performance than MMSE until error floor starts to appear. It can also be concluded that large frame size provides diversity gain over small frame size.

Now, let us look at the performance of MMSE-SIC receiver. Code block error rate (BLER) vs SNR performance is shown in Figure 4.14. Code BLER is used as the performance metric, since the MMSE-SIC takes advantage of block forward error correcting codes. In order to compare the performance of the MP algorithm [101], the soft values from MP equalizer are passed to LDPC decoder. If we look at the curves for large frame with 4-QAM, then one can note that at BLER of 10^{-2}, MP is better than MMSE by $\sim$ 1 dB. However, SIC is found to be within 0.3 dB of MP algorithm. When we shift our attention to BLER level of 10^{-3}, one can find that SIC performs almost as good as the MP algorithm. Thus, it can be concluded that gain in uncoded performance that MP has over MMSE is eroded with the use of forward error correcting codes.

For the small frame with $4 - QAM$ one may observe that SIC provides little gain over MMSE, while both are $\approx$ 1 dB worse compared with the MP algorithm for BLER upto 10^{-3}. When one considers higher reliability situations, i.e., BLER below 10^{-4}, it can be seen that appearance of error floor makes MP the worst amongst the three receiver architectures.

When we consider higher order modulation, namely $16 - QAM$ in the small frame, we can see that while MP algorithm is restricted by its error floor, SIC provides a gain of 0.5 dB over MMSE at BLER of 10^{-3}, and about 2.5 dB at higher reliability levels (lower BLER).

Thus, based on the above observations for the evaluations made in the scenarios as described above, one can state that MP shows potential as a receiver with larger gain over MMSE when the key performance indicator is uncoded BER. However, it fails to maintain the advantage when practical forward error correcting codes, which are necessary components of advanced broadband mobile communication system, are employed. Such

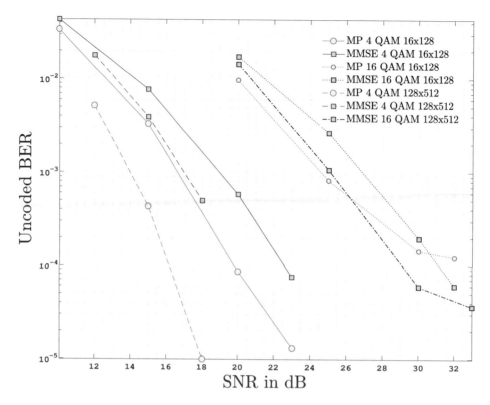

Figure 4.13   Uncoded BER in EVA channel.

observations are stronger with higher order modulation. Thus, performance evaluation for 64-QAM is made between SIC and MMSE receivers only.

It can be observed from the same figure that for 64-QAM with large frame in EVA channel that SIC provides a gain of 0.5dB over MMSE at BLER of 10^{-1}. The gain increases to more than 1 dB at BLER of 10^{-2}, and 3.5 dB at BLER of 10^{-3} and to a massive 7.5 dB at BLER of 7×10^{-4}. It can also be noted that, whereas MMSE gets limited by the appearance of error floor near BLER of 7×10^{-4}, SIC touches floor only around 10^{-5}.

The previous configuration is then evaluated in EVB channel, which has a large delay spread of $20\mu s$. It is seen that SIC provides a gain of more than 2 dB at BLER of 10^{-1} and a huge 6 dB at BLER of 3×10^{-2}. It can also be noted that while SIC easily crosses 10^{-2}, MMSE floors at a higer level.

Therefore, one can conclude that SIC provides significant benefits especially in higher order modulation and adverse channel conditions. It may be recalled that the particular realization of SIC has the unique advantage of avoiding error propagation which is otherwise a significant bottleneck for such receivers.

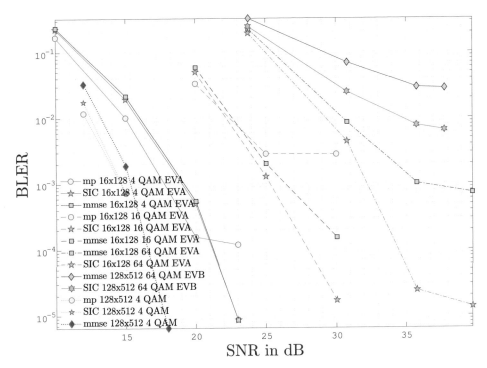

Figure 4.14 Coded BLER performance of different OTFS receivers.

4.4 Chapter Summary

In this chapter, we have explored belief propagation (BP) equalization algorithm for sparse systems.

A low complexity LMMSE receiver for OTFS is also described. It can achieve upto 10^7 times complexity reduction over direct implementation without having any performance loss.

A SIC based receiver for OTFS is also described, which is found to have comparable performance as the MP algorithm for lower modulation order while achieves significantly better performance than MP and MMSE for higher order modulations, while it has much lower complexity than the MP receiver.

Thus, it can be stated from the results discussed above that the coded SIC receiver for OTFS is a potential receiver architecture for making OTFS a feasible technology for 6G, which targets higher data rates and higher reliability at higher mobility.

5

Circulant Pulse Shaped OTFS

Conventional OTFS system discussed in the previous chapters use one symbol long rectangular pulse shape. It is known that rectangular pulse has very high out of band (OoB) radiation. A linear pulse reduces spectral efficiency as it spreads out of the symbol in frequency domain. Such spreading of signal across the frequency band is undesirable in various scenarios such as multi-user communication in adjacent frequency bands/channels. Instead of using rectangular pulse shape, use of frequency localized pulse shapes are suggested [110] in order to reduce OoB radiation in multi-carrier communication. This motivates us to consider circular pulse shape in order to retain spectral efficiency of OTFS system. Thus, a circular pulse shaped (CPS) framework for OTFS in which pulses are circular and can span the whole frame duration is considered.

5.1 Chapter Outline

CPS-OTFS system is introduced in section 5.2 which is followed by circulant Dirichlet pulse shaped OTFS system in section 5.4. Low complexity transmitter is presented in section 5.3. Section 5.6 presents performance analysis of CPS-OTFS. Chapter summary is presented in section 5.7.

5.2 Circular Pulse Shaped OTFS (CPS-OTFS)

CPS-OTFS system can be understood in the light of Figure 5.1. We consider an OTFS system with $T_f = NT$ frame duration and $B = M\Delta f$ bandwidth having N number of time symbols with T symbol duration and M number of subcarriers with Δf bandwidth. Moreover, we assume that OTFS system is critically sampled i.e., $T\Delta f = 1$.

Each TF domain data $X(n,m)$ is pulse shaped using circular pulse shape $g_{(n,m)}(t) = g(t - mT)_{T_f} e^{j2\pi n\Delta f(t-mT)}$, where $g(t)$ a prototype pulse shape having length T_f. Time

129

Figure 5.1   Circular pulse shaped OTFS system.

domain transmitted signal is obtained as,

$$s(t) = \sum_{n=0}^{N-1} \sum_{m=0}^{M-1} X(n,m)g((t-mT) \mod T_f)e^{j2\pi n \Delta f(t-mT)} \tag{5.1}$$

CPS-OTFS can also be viewed as an ***ISFFT precoded GFDM*** [74] system where TF modulation is performed as GFDM modulation. CPS-OTFS system converges to pulse shaped OTFS in [92] when,

$$g(t) = \begin{cases} a(t) \text{ if } 0 < t < T \\ 0 \text{ otherwise} \end{cases}, \tag{5.2}$$

where, $a(t)$ is a T duration pulse shape. rectangular pulse shaped OTFS (RPS-OTFS) as in [91, 99 and 101] can be obtained by setting $a(t) = 1$. After taking samples at $\frac{T}{M}$ sampling duration, transmitted signal is given as,

$$s(r) = \sum_{n=0}^{N-1} \sum_{m=0}^{M-1} X(n,m)g((r-nM) \mod MN)e^{j2\pi \frac{mr}{M}}, \tag{5.3}$$

for $r = \mathbb{N}[0 \ MN - 1]$.

We collect values of $d(k,l)$ in a vector as $\mathbf{d} = [d(0,0) \ d(0,1) \ \cdots d(0,M-1) \ d(1,0) \ d(1,1) \cdots d(1,M-1) \cdots d(N-1,0) \ d(N-1,1) \cdots d(N-1,M-1)]^T$ and similarly, we take values of $X(n,m)$ is a vector as, $\mathbf{x} = [X(0,0) \ X(0,1) \ \cdots X(0,M-1) \ X(1,0) \ X(1,1) \cdots X(1,M-1) \cdots X(N-1,0) \ X(N-1,1) \cdots X(N-1,M-1)]^T$. Using (5.1), $\mathbf{x}$ is given as,

$$\mathbf{x} = \underbrace{\mathbf{U}_M^H \mathbf{P} \mathbf{U}_N}_{\mathbf{A}_{DD}} \mathbf{d}, \tag{5.4}$$

where, $\mathbf{U}_N = \mathbf{I}_M \otimes \mathbf{W}_N$, $\mathbf{U}_M = \mathbf{I}_N \otimes \mathbf{W}_M$ and $\mathbf{P}$ is a permutation matrix, whose elements can be given as,

$$\mathbf{P}(s,q) = \begin{cases} 1 \text{ if } q = (s \mod M)N + \lfloor \frac{s}{M} \rfloor \\ 0 \text{ Otherwise.} \end{cases}, \tag{5.5}$$

for $q, s \in \mathbb{N}[0 \; MN - 1]$. Further, collecting samples $s(r)$ as $\mathbf{s} = [s(0) \; s(1) \cdots s(MN - 1)]$, $\mathbf{s}$ can be given as, $\mathbf{s} = \mathbf{A}_g \mathbf{x}$, where $\mathbf{A}_g$ is GFDM modulation matrix. Thus, transmitted signal can be given in matrix-vector form as,

$$\mathbf{s} = \mathbf{A}\mathbf{d}, \tag{5.6}$$

where,

$$\mathbf{A} = \mathbf{A}_g \mathbf{A}_{\text{DD}} \tag{5.7}$$

is CPS-OTFS modulation matrix. A cyclic prefix (CP) of length $\alpha' \geq \alpha - 1$ is appended at the end and beginning of $s(r)$, α is channel delay length i.e., $\mathbf{s}_{cp} = [\mathbf{s}(MN - \alpha' + 1 : MN - 1) \; \mathbf{s}^T \; \mathbf{s}(0 : \alpha' - 1)]$. Finally, $\mathbf{s}_{cp}$ is multiplied to a window of length $MN + 2\alpha'$ which has values of one for middle MN samples and soft edges for initial and last α' samples.

The receiver can be realized following (3.23).

5.3 Low Complexity Transmitter for CPS-OTFS

CPS-OTFS modulation using (5.6) requires complex multiplications (CMs) in the $O(M^2 N^2)$ which can be computationally a burden when values of MN are high. Here, we present a low complexity transmitter for practical implementation of CPS-OTFS. Using the factorization of $\mathbf{A}_g$, (5.6) can be given as,

$$\mathbf{s} = \underbrace{\mathbf{P}\mathbf{U}_N \mathbf{D}\mathbf{U}_N^{\mathrm{H}} \mathbf{P}^{\mathrm{T}} \mathbf{U}_M}_{\mathbf{A}_g} \underbrace{\mathbf{U}_M^{\mathrm{H}} \mathbf{P}\mathbf{U}_N}_{\mathbf{A}_{\text{DD}}} \mathbf{d} \tag{5.8}$$

$$= \underbrace{\mathbf{P}\mathbf{U}_N \mathbf{D}}_{\mathbf{A}} \mathbf{d}, \tag{5.9}$$

where, $\mathbf{D} = diag\{\lambda(0), \; \lambda(1) \cdots \lambda(MN - 1)\}$ is diagonal matrix, whose r^{th} element can be given as,

$$\lambda(r) = \sum_{m=0}^{N-1} g[mM + \lfloor \frac{r}{N} \rfloor] e^{j2\pi \frac{m(r \mod N)}{M}} \tag{5.10}$$

Thus, using (5.9) and assuming that $\mathbf{D}$ is computed offline, CPS-OTFS can be implemented using M number of N-point IFFTs and MN-point scalar multiplier. The transmitter for CPS-OTFS requires $MN + \frac{MN}{2} \log_2 N$ CMs which needs only MN more CMs than RPS-OTFS in [99]. Thus, the transmitter has $O(MN \log_2(N))$ complexity. Block OFDM requires complexity of $O(MN \log_2(M))$. Generally, $N < M$ which makes CPS-OTFS transmitter simpler than block-OFDM. It is also interesting to see that CPS-OTFS is simpler than GFDM transmitter, which requires $O(MN \log_2(MN))$ complexity. Table 5.1 presents number of complex multiplications (CMs) required for different transmitter implementation. It can be observed that OTFS has similar complexity as OFDM and 2-4 times simpler than GFDM.

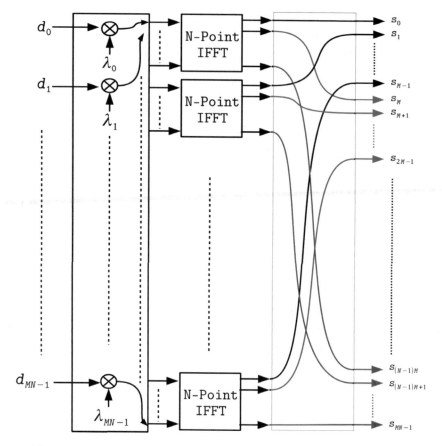

Figure 5.2 Our proposed low-complexity CPS-OTFS transmitter.

Table 5.1 Number of CMs required for various transmitters (for $\Delta f = 15$ KHz).

Waveform	Short frame ($T_f \approx 1.1ms$) $N = 16$ and $M = 512$	Long frame ($T_f \approx 8.5ms$) $N = 128$ and $M = 512$
OFDM	36864	294912
GFDM	77824	819200
RPS-OTFS	16384	229376
CPS-OTFS	24576	294912

5.4 Circular Dirichlet Pulse Shaped OTFS (CDPS-OTFS)

For unitary $\mathbf{A}$ in (5.9), $\mathbf{D}$ should be unitary as $\mathbf{PU}_N$ are unitary i.e., $\mathbf{DD}^{\mathrm{H}} = \mathbf{I}_{MN}$ or $abs\{\mathbf{D}\} = \mathbf{I}_{MN}$. Trivial solution to this is a rectangular pulse for which $\mathbf{D} = \mathbf{I}_{MN}$. A class of prototype pulses for which $abs\{\mathbf{D}\} = \mathbf{I}_{MN}$ are called constant magnitude characteristics matrix (CMCM) pulses [66]. Drichlet pulses are CMCM pulses which can be obtained by taking inverse Fourier transform of rectangular pulse. Let $\mathbf{g}_{rect} = \frac{1}{\sqrt{M}}[1\ 1 \cdots 1\ \mathbf{0}_{MN-M}]_{MN}^{\mathrm{T}}$ be a rectangular pulse, Dirichlet pulse can be found by

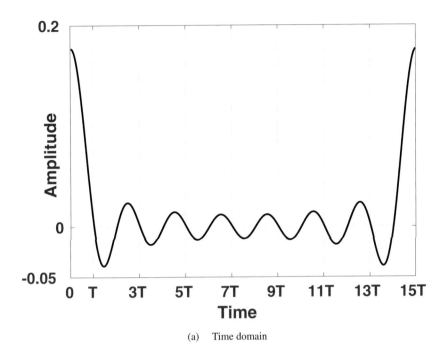

(a)   Time domain

(b)   Frequency domain

Figure 5.3   *Continued*

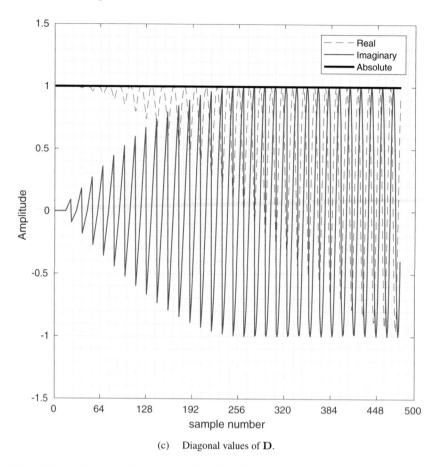

(c)   Diagonal values of **D**.

Figure 5.3 Illustration of frequency localization of Dirichlet pulse for $N = 15$ and $M = 32$.

$\mathbf{g} = \mathbf{W}_{MN}^{\mathrm{H}} \mathbf{g}_{rect}$ whose time and frequency domain shape can be understood by Figure 5.3a and 5.3b, respectively. As opposed to rectangular pulses, Drichlet pulses are localized in frequency which helps in reducing OoB radiation. Figure 5.3c shows the diagonal values of **D**. It can be observed that all the diagonal values $abs\{\mathbf{D}\}$ are ones. This reconfirms that **A** is unitary for CDPS-OTFS.

5.5 Remarks on Receiver Complexity

Since, **A** for RPS-OTFS as well as CDPS-OTFS is unitary, (3.23) is simplified to

$$\hat{\mathbf{d}} = \underbrace{\mathbf{A}^{\mathrm{H}}}_{MF} \underbrace{\mathbf{H}^{\mathrm{H}}[\mathbf{H}\mathbf{H}^{\mathrm{H}} + \frac{\sigma_{\nu}^{2}}{\sigma_{d}^{2}}\mathbf{I}]^{-1}}_{MMSE-CE} \mathbf{r}. \qquad (5.11)$$

Thus, the MMSE receiver for CDPS-OTFS just like that of RPS-OTFS (as discussed in previous chapter) is simplifed to a two-stage receiver. The first stage is MMSE channel equalization (MMSE-CE), followed by a matched filter (MF) for OTFS. It is straightforward to see that MF operation for CDPS-OTFS requires additional MN-point scalar multiplications for multiplication of diagonal values of $\mathbf{D}^H$ as compared to RPS-OTFS. It can be concluded that implementation of MMSE receiver for CDPS-OTFS is only slightly higher than that of MMSE receiver for RPS-OTFS.

5.5.1 LMMSE Receiver for GFDM and OFDM over TVC

Low complexity receiver discussed for CDPS-OTFS can easily be extended to GFDM and OFDM by setting $\mathbf{A} = \mathbf{P}\mathbf{U}_N\mathbf{D}\mathbf{U}_N^H\mathbf{P}^T\mathbf{U}_M$ for GFDM and $\mathbf{A} = \mathbf{I}_N \otimes \mathbf{W}_M$ for OFDM. To do so, $\mathbf{r}_{ce} = \mathbf{H}_{eq}\mathbf{r}$ is performed by computing (4.21-4.27) as discussed in Sec. 4.2.2.2 and 4.2.2.3. For GFDM, $\hat{\mathbf{d}} = (\mathbf{P}\mathbf{U}_N\mathbf{D}\mathbf{U}_N^H\mathbf{P}^T\mathbf{U}_M)^\dagger\mathbf{r}_{ce}$ can be computed using N number of M-point FFTs, $2M$ number of N-point FFTs and MN number of complex scalar multiplier.

5.6 Simulation Results

The power spectral density (PSD) for different waveforms is plotted in Figure 5.4. The legend OFDM refers to block OFDM having M subcarriers and N time slots. It can be observed that RPS-OTFS has PSD similar to OFDM. Whereas, it is observable that GFDM and CDPS-OTFS have almost identical PSD. It can also be observed that the side lobes of RPS-OTFS vary between -30 dB to -50 dB. On the other hand it is seen that CDPS-OTFS and GFDM both have much lower side-lobes. It is also notable that at the edge subcarriers, both GFDM and CPS-OTFS have nearly 50 dB lower OoB than RPS-OTFS. Thus it can be concluded that GFDM and CDPS-OTFS both have very low OoB spectral leakage.

A plot of complementary cumulative density function (CCDF) of PAPR for different waveforms is shown in Figure 5.5. It can be observed that the PAR of RPS-OTFS and OFDM are very close to each other. On the other hand it is found that Dirichlet pulse shaping reduces the PAPR as compared to rectangular pulse shaping in OTFS. It can be seen that CDPS-OTFS has about 1.6 dB gain over RPS-OTFS. It can also be observed that GFDM has PAPR values similar to CDPS-OTFS.

Table 5.2   Simulation parameters.

Number of subcarriers M	512
Number of time slots N	127
Window	Mayer root raised cosine with roll of factor =1 [46]
Mapping	4 QAM
Sub-carrier bandwidth	15 KHz
Channel	Extended vehicular-A (EVA) [30]
Vehicular speed (in Kmph)	500
Carrier speed	4 GHz
Gaurd (null) subcarriers for OoB computation	$\mathbb{N}[1\ 128] \bigcup \mathbb{N}[384\ 512]$
Circular prefix (CP) or postfix value, α'	64

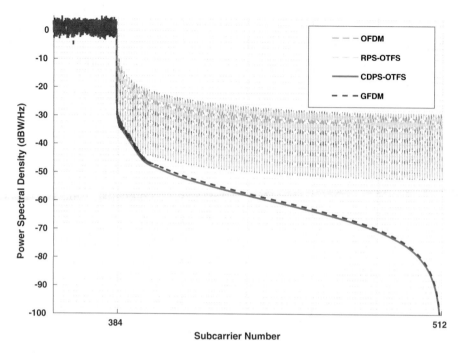

Figure 5.4   Comparison of out of band leakage for different waveforms.

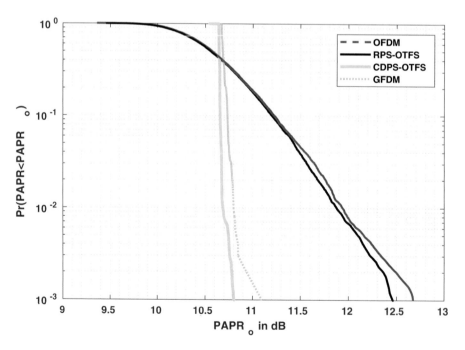

Figure 5.5   PAPR comparison of RPS-OTFS and CDPS-OTFS.

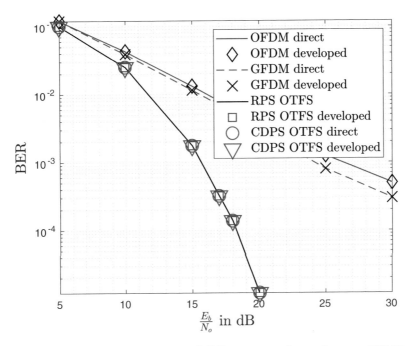

Figure 5.6 The BER performance of different waveforms for $v = 500$ Kmph.

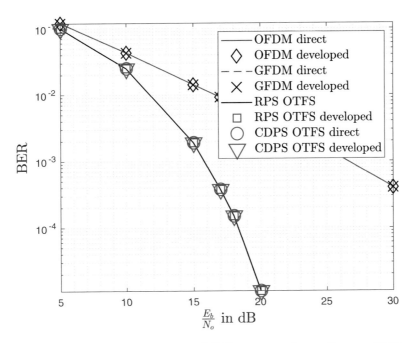

Figure 5.7 The BER performance of different waveforms for $v = 30$ Kmph.

The BER performance of CDPS-OTFS in EVA channel is shown in Figure 5.6 and 5.7 for 500 Kmph and 30 Kmph vehicular speed respectively. Direct implementation of the receivers are obtained by using (5.11). It can be observed that the low complexity receivers do not incur performance loss when compared with direct implementation of receivers. It can also be seen that the BER performance of CDPS-OTFS is similar to that of RPS-OTFS. This can be ascribed to the unitary property of CDPS-OTFS. Further, both RPS-OTFS and CDPS-OTFS have diversity gain of more than 10 dB over OFDM or GFDM at the BER of 10^{-3}, which is aligned with results in [101, 91, 111].

5.7 Chapter Summary

In this chapter, a circular Dirichlet pulse shaped OTFS (CDPS-OTFS) is described. It is found that such pulse shape can reduce OoB radiation by about 50 dB over rectangular pulse shaped OTFS (RPS-OTFS) system. CDPS-OTFS is also shown to have more than 1 dB lesser PAPR over the RPS-OTFS system. It is also shown that a low complexity transmitter to implement CDPS-OTFS has similar computational complexity to that of the OFDM system.

6

Channel Estimation in OTFS

In this chapter we discuss two channel estimation techniques for OTFS, namely delay-Doppler domain channel estimation and time domain channel estimation.

6.1 Delay Doppler Channel Estimation

Till now, we have seen the system model of OTFS. We have also understood how OTFS and OFDM signals can be generated in the same flow. We have also seen different receiver structures for OTFS. The receiver implementation and its performance was shown with ideal channel knowledge (estimates). However, in realistic situations ideal channel coefficients are not available. One has to estimate channel coefficients using pilots and use the estimated channel coefficients for channel equalization. In this chapter, we explore channel estimation techniques in different domains, namely: (1) delay Doppler domain (2) time domain. First we detail the delay Doppler domain channel estimation. In this part we also compare the performance of OTFS against variable subcarrier bandwidth (VSB) OFDM, the one used to create resource element in 5G-NR physical layer [112], where VSB reconfigurability is expressed in terms of "numerology".

We also delve into pilot power aspect, since the pilot signal in OTFS is an impulse in De-Do domain.

We also describe a time domain channel estimation for OTFS. It is shown that the time domain channel estimation, when used along with time domain channel equalization is capable to addressing residual time and frequency synchronization errors.

6.1.1 Pilot Structure

OTFS system followed here is as described in Chapter 3. We consider RCP-OTFS system operating with TF resource of total T_f sec. duration and B Hz. The bandwidth is divided into M number of subcarriers having Δf sub-carrier bandwidth and we transmit N number of symbols having T symbol duration, thus $B = M\Delta f$ and $T_f = NT$. Furthurmore, OTFS is critically sampled, i.e., $T\Delta f = 1$

The source bitstream is encoded using LDPC codes and then passed through symbol mapper. The QAM modulated Do-De data and pilot symbols are arranged over Do-De lattice $\Lambda = \{(\frac{k}{NT}, \frac{l}{M\Delta f})\}$, $k \in \mathbb{N}[0 \ N - 1]$, $l \in \mathbb{N}[0 \ M - 1]$ as shown in Figure 6.1. Doppler-delay signal can be given as,

$$
x(k, l) = \begin{cases}
x_p, & k = K_p \ \& \ l = L_p \\
0 & K_p - 2k_\nu \leq k \leq K_p + 2k_\nu \ \& \\
& L_p - 2l_\tau \leq l \leq L_p + 2l_\tau \\
d(k, l), & otherwise
\end{cases} \tag{6.1}
$$

where, $d(k, l) \in \mathbb{C}$ is the QAM data symbol, $x_p = \sqrt{(P_{pilot})}$ is the pilot symbol, P_{pilot} is power of pilot symbol, $k_\nu = \lceil \nu_{max}NT \rceil$ and $l_\tau = \lceil \tau_{max}M\Delta f \rceil$ are maximum doppler length and delay length of the channel[113] and (K_p, L_p) is the pilot location, $K_p \in \mathbb{N}[2k_\nu + 1 \ N - 2k_\nu - 2]$, $L_p \in \mathbb{N}[l_\tau + 1 \ M - l_\tau - 2]$.

Then, $x(k, l)$ is mapped to TF data $Z(n, m)$ on lattice $\Lambda^\perp = \{(nT, \ m\Delta f)\}$, $n \in \mathbb{N}[0 \ N - 1]$ and $m \in \mathbb{N}[0 \ M - 1]$ by using inverse symplectic finite Fourier transform (ISFFT).

6.1.2 Delay-Doppler Channel Estimation

Channel matrix $\mathbf{H}$ is estimated from the pilot symbols. The received signal is transformed to Do-De domain as,

$$
\mathbf{y} = \mathbf{A}^\dagger \mathbf{r} \tag{6.2}
$$

The vector $\mathbf{y}$ is reshaped back to the $N \times M$ grid as,

$$
\mathbf{Y}(k, l) = \mathbf{y}(k + lN) \tag{6.3}
$$

where, $k \in \mathbb{N}[0 \ N - 1]$, $l \in \mathbb{N}[0 \ M - 1]$. The channel estimation from here proceeds according to [98], in which the the nonzero pilot at location (K_p, L_p) as shown in Figure 6.1 is spread to locations (k, l), $k \in \mathbb{N}[K_p - k_\nu \ K_p + k_\nu]$ and $l \in \mathbb{N}[L_p \ L_p + l_\tau]$ because of the channel.

The channel parameters (h_p, k_p, l_p) are extracted from this region using the threshold based scheme with threshold(Υ) as $3\sigma_\mathbf{n}$ given as,

$$
(\hat{h}_p, \hat{k}_p, \hat{l}_p) = (Y(k, l), k, l)
$$

$\ni \|Y(k, l)\| > \Upsilon \ \forall \ K_p - k_\nu \leq k \leq K_p + k_\nu \ \& \ L_p \leq l \leq L_p + l_\tau$

From the channel parameters, the channel matrix $\mathbf{H}$ is formed as,

$$
\hat{\mathbf{H}} = \sum_{p=1}^{\hat{P}} \hat{h}_p \Pi^{\hat{l}_p} \Delta^{\hat{k}_p}
$$

where $\hat{P}$ is the number of taps detected.

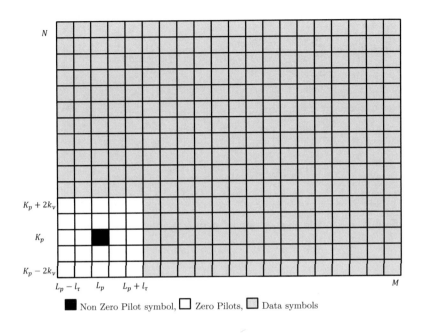

Figure 6.1   RCP-OTFS.

6.1.3 Channel Equalization

The Do-De data is estimated by equalizing the received time domain signal using MMSE [94] as,

$$\hat{\mathbf{x}} = \mathbf{H_{mmse}}\mathbf{r} \tag{6.4}$$

where,

$$\mathbf{H_{mmse}} = (\hat{\mathbf{H}}\mathbf{A})^{\dagger}[(\hat{\mathbf{H}}\mathbf{A})(\hat{\mathbf{H}}\mathbf{A})^{\dagger} + \frac{1}{\Gamma}\mathbf{I}]^{-1}$$

To decode the equalized data using LDPC decoder, the log-likelihood ratios (LLRs) are passed to the decoder which are calculated from the equalized symbols as described in Section 4.3.2 but for $i = 1$ only.

6.1.4 Performance of Channel Estimation

To evaluate the performance of OTFS system, the system depicted in Figure 6.2 is simulated. It is slightly different than the one used in system model development. This is due to consideration of practical FECs and channel estimation.

In Figure 6.3, the BLER performance of the RCP-OTFS with channel estimation for 16QAM modukation in rural macro(RMa) channel model at user speed of 500 Kmph provided by 3GPP is shown. It can be observed that there is almost no significant difference between performance using ideal channel estimates and real channel estimates.

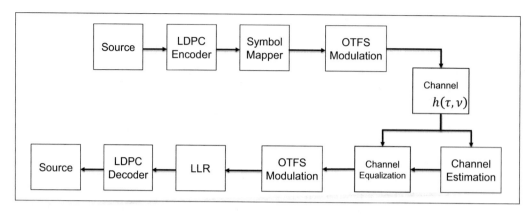

Figure 6.2   Block diagram of simulated system model.

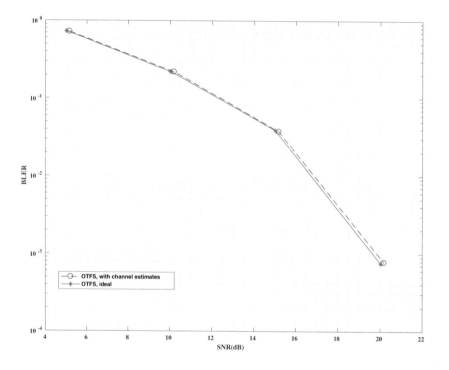

Figure 6.3   BLER results for 16QAM with channel estimation.

6.1.5 VSB OFDM Overview

We compare the performance of VSB OFDM as it is known for quite some time now that (VSB-OFDM)[13],[9], [114], and [14] can be used to improve the resilience of OFDM to ICI as done in 5G-NR, whereas OTFS claims to improve upon 5G NR. The performance comparison is made in realistic channel profile as specified in [115] with LDPC codes and practical channel estimation.

VSB-OFDM system with total frame duration T_f sec. and bandwidth B Hz is considered. Base subcarrier bandwidth is denoted with Δf Hz and variability in the subcarrier bandwidth is introduced by the parameter μ. For a given μ, the subcarrier bandwidth is $2^\mu \Delta f$ Hz. We have total $2^{-\mu}M$ number of subcarriers and $2^\mu N$ number of OFDM symbols having $2^{-\mu}T$ symbol duration , thus, $B = M\Delta f$ and $T_f = NT$.

6.1.5.1 Transmitter

The physical resource block (PRB) frame structure of 5G NR is used for VSB-OFDM system. Each PRB consists of 12 subcarriers $\times$ 14 time slots with 8 reference signals (RS) as shown in Figure 6.4. Then the PRBs are arranged to fill the TF grid of B Hz and T_f sec. The number of PRBs in a given frame is,

$$N_{PRB} = \left\lfloor \frac{B}{12.2^\mu.\Delta f} \right\rfloor \left\lfloor \frac{T_f}{14.2^{-\mu}.T} \right\rfloor$$

The source bits are encoded using LDPC code and then passed through the symbol mapper. The modulated QAM symbols are arranged in the N_{PRB} number of PRBs to form the TF signal $\acute{x}(m,n)$, $m \in \mathbb{N}[0 \;\; 2^{-\mu}M - 1]$ and $n \in \mathbb{N}[0 \;\; 2^\mu N - 1]$ in a frame. Then, the

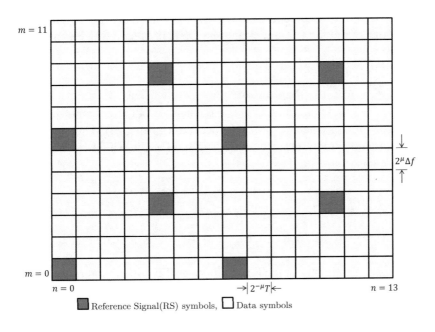

Figure 6.4 VSB-OFDM.

nth time domain OFDM symbol can be given as,

$$\acute{s}_n(t) = \sum_{m=0}^{2^{-\mu}M-1} \acute{x}(m,n)e^{j2\pi m2^{\mu}\Delta f(t-n2^{-\mu}T)} \tag{6.5}$$

where, $2^{-\mu}T$ is the one OFDM symbol duration without CP. Further, if $\acute{\mathbf{x}}_\mathbf{n}$ represents the discrete version of $\acute{s}_n(t)$ sampled at $\frac{1}{B}$, then the $2^{\mu}N$ concatenated OFDM symbols can be given as,

$$[\acute{\mathbf{x}}_\mathbf{0}\ \acute{\mathbf{x}}_\mathbf{1}\ \cdots\ \acute{\mathbf{x}}_{\mathbf{2}^{\mu}\mathbf{N}}] = \mathbf{W}_{\mathbf{2}^{-\mu}\mathbf{M}}\acute{\mathbf{X}} \tag{6.6}$$

where, $\acute{\mathbf{X}}$ is the TF frame given as,

$$\acute{\mathbf{X}} = \begin{bmatrix} \acute{x}(0,0) & \cdots & \acute{x}(0,2^{\mu}N-1) \\ \acute{x}(1,0) & \cdots & \acute{x}(1,2^{\mu}N-1) \\ \vdots & \ddots & \vdots \\ \acute{x}(2^{-\mu}M-1,0) & \cdots & \acute{x}(2^{-\mu}M-1,2^{\mu}N-1) \end{bmatrix}$$

Then, Cyclic Prefix (CP) is appended to each OFDM symbol of duration $2^{-\mu}T_{cp}$ sec such that $T_o = T + T_{cp}$.

6.1.5.2 Receiver

Assuming the CP duration is greater than maximum excess delay of channel, one OFDM symbol duration is within a coherence time and subcarrier bandwidth within coherence bandwidth. Then, after removing CP and taking the DFT, the received TF symbol $\acute{y}(m,n)$ can be expressed as,

$$\acute{y}(m,n) = \acute{h}(m,n)\acute{x}(m,n) + \acute{v}(m,n) \tag{6.7}$$

where, $m \in \mathbb{N}[0\ \ 2^{-\mu}M-1]$ is the subcarrier index and $n \in \mathbb{N}[0\ \ 2^{\mu}N-1]$ is the OFDM symbol index. $\acute{h}(m,n) \in \mathbb{C}$ defined in (3.19) is the channel coefficient and $\acute{v}(m,n) \in \mathbb{C}$ is white Guassian noise with variance σ_V^2 at mth subcarrier of nth OFDM symbol.

For channel estimation, the reference signal symbols in the a PRB is used to get the estimates at the pilot location using MMSE estimation as,

$$\hat{h}(m_{RS}, n_{RS}) = \frac{\acute{x}(m_{RS}, n_{RS})^{\dagger}\ \acute{y}(m_{RS}, n_{RS})}{\|\acute{x}(m_{RS}, n_{RS})\|^2 + \sigma_V^2} \tag{6.8}$$

where, $(m_{RS}, n_{RS}) = \{(0,0),(6,0),(3,4),(9,4),(0,7),(6,7),(3,11),(9,11)\}$ are the position of RS in a PRB.

The obtained channel estimates at the RS locations in TF grid are interpolated to get the channel estimates at the data locations using DFT interpolation along frequency axis and linear interpolation along time axis as in [116].

The estimate of data symbols are obtained at the receiver using,

$$\hat{x}(m,n) = \frac{\acute{y}(k,l)}{\hat{h}(k,l)} \tag{6.9}$$

The estimated symbols are used to generate the channel LLR values of bits corresponding to the symbol by substituting $\sigma^2 = diag\{vec\{\sigma^2_{\mathbf{V}_{\text{eff}}}\}\}$ in (3.5) where $\sigma^2_{V_{eff}}(m,n) = \frac{\sigma^2_V}{\|\hat{h}(m,n)\|^2}$. Then, LLRs are used to recover the data as described in 3.5.2.

6.1.6 Pilot Power in OTFS and VSB-OFDM

Pilots are needed for channel estimation in all transmission schemes. In this section, we describe the pilot structure to be used in OTFS as in [98]. We also describe the pilot structure used for evaluating VSB-OFDM. In the performance evaluation, we intend to keep same total transmit power for OTFS and OFDM. As shown in Figure 6.4, there are 8 pilots per PRB in VSB-OFDM system considered. The total number of pilots in VSB-OFDM frame is $N_{p,ofdm} = 8N_{PRB}$, while total number of pilots in RCP-OTFS is $N_{p,otfs} = (4k_\nu+1)(2l_\tau+1) - 1$[98]. If we let the total pilot power to be equal, i.e., $P_{pilot} = n_{p,ofdm}P_T = n_{p,otfs}P_T$, where P_T is the transmit power, $n_{p,otfs}$ and $n_{p,ofdm}$ are the ratio of number of pilot symbols to total number of symbols for RCP-OTFS and VSB-OFDM, respectively. Since, RCP-OTFS uses only one nonzero pilot at location (K_p, L_p) and $(N_{p,otfs} - 1)$ zero pilots as shown in Figure 6.1, total pilot power is placed on the pilot symbol (x_p) which results in an uneven power distribution for pilots and data in RCP-OTFS given by $\Delta P = 10 log_{10}(\frac{P_{pilot}}{P_{data}})$. Value of ΔP is 34 dB for the RCP-OTFS system, for parameters given in Table 6.2. If P_T is to be kept same as VSB-OTFS system. We evaluate the impact of ΔP on PAPR in Sec. 6.1.7 below.

6.1.7 Results

The performance comparison of LDPC coded performance of block OFDM, OTFS and VSB-OFDM system. Table 6.2 describes the simulation parameters used in the evaluation.

Table 6.1 Simulation parameters.

Parameter	Value
Carrier frequency(f_c)	6 GHz
Bandwidth(B)	7.68 MHz
Frame time(T_f)	10 ms
Subcarrier bandwidth(Δf)	15 KHz
RCP-OTFS parameters	M=512, N=128, K_p=80, L_p=16
VSB-OFDM parameters	M=512, N=128, μ=0,1,2,3
Equivalent OFDM parameters	M_{bofdm}=128, N_{bofdm}=512, Δf_{bofdm} = 60 KHz
Channel Model[115]	TDL-A, DS=37 ns, Rural Macro
CP duration(T_{cp})	4.69 μs
UE speed	500 Kmph
FEC	QC-LDPC, coderate 2/3, codeword length 1944

The key performance indicator (KPI) used is block error rate (BLER), where block indicates a code block of LDPC code.

The frame size for block OFDM is considered the same frame size as RCP-OTFS. TF slots for block OFDM can be evaluated following Sec. 3.3 as $M_{bofdm} = 128$, $N_{bofdm} = 512$, and $\Delta f_{bofdm} = 60$ KHz. *It may be noted that the same LMMSE equalizer is used for both block OFDM and RCP-OTFS, so that any difference in BLER performance can be attributed to interleaving only.* The SNR due to CP, which is $10log_{10}(\frac{[T_{cp}B]}{M}) = 0.3$ dB for VSB-OFDM and $10log_{10}(\frac{[T_{cp}B]}{MN}) = 0.0023$ dB for OTFS is also adjusted in the results. Identical transmit power is used for all schemes. CP length and Doppler generation is described in earlier chapters.

We consider that a large ΔP may have an effect on PAPR of s (3.10). Therefore, we analyze its effects as well. We present the CCDF of RCP-OTFS for $\Delta P = 34$ dB and $\Delta P = 28$ dB in Figure 6.5. If one considers same P_T for both RCP-OTFS and VSB-OFDM then $\Delta P = 34$ dB. RCP-OTFS is seen to have nearly 3.5 dB higher PAPR than that

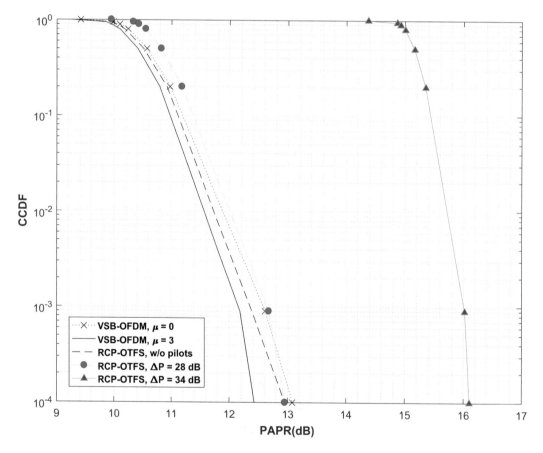

Figure 6.5   CCDF of PAPR for OFDM and OTFS.

of VSB-OFDM ($\mu = 0$). The value of ΔP is reduced to 28 dB, so that PAPR of OTFS may be reduced. It can be seen from the curves that with the reduced ΔP, the PAPR of OTFS turns to be almost same as that of OFDM. It may also be noted that for the configuration under consideration, if ΔP is kept ≤ 28 dB, then there is no further reduction in PAPR of RCP-OTFS signal. The reduction in ΔP implies that the power of data symbols in RCP-OTFS is higher than that of VSB-OFDM by 0.154 dB/symbol. This leads to an additional SNR gain for OTFS.

The different waveforms, namely (VSB-OFDM, RCP-OTFS and its equivalent block OFDM (section 3.3)) are evaluated for 16-QAM modulation at the vehicular speed of 500 Kmph and results are presented in Figure 6.6. It can be observed that VSB-OFDM performs better with increasing μ. This is because the effect of ICI due to Doppler reduces with increased in subcarrier bandwidth. It was shown in [9] that post processing SNR (Γ) for OFDM under Doppler, $\propto (\frac{\Delta f}{\nu_{max}})^2$. It can also be observed that OTFS outperforms VSB-OFDM even with $\mu = 3$, the highest permissible value of μ as per 5G-NR.

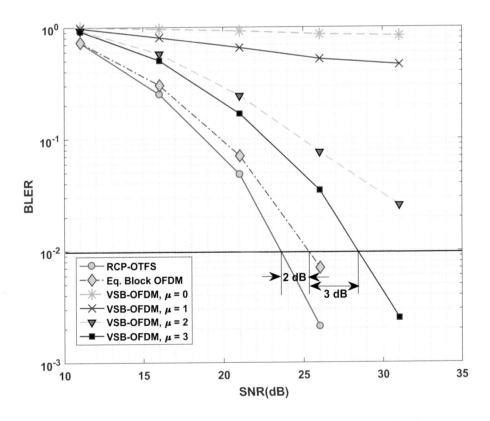

Figure 6.6   BLER performance for 16 QAM at 500 Kmph.

OTFS has 3.5 dB SNR advantage over VSB-OFDM ($\mu = 3$) at BLER of 10^{-1}. At BLER of 10^{-2} this gain further increases to ~ 5 dB. This result can be attributed to two reasons: (i) OTFS has an interleaving gain and, (ii) OTFS uses ICI cancellation LMMSE receiver, whereas VSB-OFDM uses a simple single-tap receiver.

Continuing to block OFDM, it can be seen that block OFDM, with LMMSE interference cancellation receiver, has a SNR gain of around 2.5 dB and 3 dB at BLER 10^{-1} and 10^{-2}, respectively over VSB-OFDM with single tap equalizer. At the BLER of 10^{-1} and 10^{-2}, OTFS has gain of 1 dB and 2 dB, respectively over block OFDM. Now, if we carefully analyze the results while considering that both block OFDM and RCP-OTFS use LMMSE equalizer, whereas the difference in transmitted signal is time interleaving as explained in Sec. 3.3 and shown in Figure 3.3 and Figure 3.4, then one can attribute this gain of OTFS over block OFDM to interleaving.

Thus, it can be concluded that OTFS differs from OFDM only in interleaving, which provides it an SNR gain over its sibling.

6.2 Time Domain Channel and Equalization

In this section, we describe time domain channel estimation which also captures the effect of residual synchronization errors. The estimated channel coefficients are used to time domain channel equalisation.

6.2.1 System Model

Here, we consider a CP-OTFS system with M subcarriers, each of Δf Hz bandwidth, N symbols of duration $T_u = \frac{1}{\Delta f}$ sec. each with T_{CP} sec. long CP. The system has bandwidth $B = M\Delta f$ Hz. and total frame duration $T_f = NT$ sec., where, $T = T_u + T_{CP}$.

6.2.1.1 Transmitter
After appending CP to the baseband signal in (3.2) we get,

$$s(t) = \frac{1}{\sqrt{M}} \sum_{n=0}^{N-1} \sum_{m=0}^{M-1} X(m,n)g(t-nT)e^{j2\pi m\Delta f(t-T_{CP}-nT)}, \qquad (6.10)$$

where, $g(t) = 1$ if $0 \leq t \leq T$, and $g(t) = 0$, otherwise. The baseband signal $s(t)$ is upconverted to the RF carrier frequency f_c to obtain the RF signal $s_{RF}(t) = s(t)e^{j2\pi f_c t}$.

The channel used is described in Section 3.5. The RF equivalent channel can be given as, $h_{RF}(\tau, \nu) = h(\tau, \nu)e^{j2\pi f_c \tau}$.

The received signal can be written as, $r_{RF}(t) = \int_{\tau=0}^{T_{max}} \int_{\nu=-\nu_{max}}^{\nu_{max}} \Big(h_{RF}(\tau, \nu)s_{RF}(t-\tau)$ $e^{j2\pi\nu(t-\tau)} \Big) d\nu d\tau + v_{RF}(t)$, where, $v_{RF}(t)$ is Gaussian noise with variance σ_v^2. Therefore, $r_{RF}(t) = \sum_{p=1}^{P} \Big(h_p e^{j2\pi f_c \tau_p} s(t - \tau_p)e^{j2\pi f_c(t-\tau_p)}e^{j2\pi\nu_p(t-\tau_p)} \Big) + v_{RF}(t)$. The received signal after downconversion to base band is $r(t) = r_{RF}(t)e^{-j2\pi f_c' t}$, where $f_c' = f_c - \delta f_c$ is

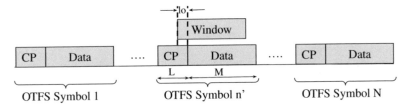

Figure 6.7 FFT window matching.

the receiver carrier frequency with offset δf_c. The signal $r(t)$ is,

$$r(t) = \sum_{p=1}^{P} h_p e^{j2\pi\delta f_c \tau_p} s(t - \tau_p) e^{j2\pi(\delta f_c + \nu_p)(t - \tau_p)}. \tag{6.11}$$

The signal $r(t)$ sampled at $Fs = B = 1/T_s = (M + L)/T$, where $L = \lceil T_{CP}B \rceil$ is the length of sampled CP and $T = T_u + T_{CP} = (M + L)T_s$, becomes,

$$r(l') = r(l'T_s)$$
$$= \sum_{p=1}^{P} h_p e^{j2\pi\delta f_c \tau_p} \left(\frac{1}{\sqrt{M}} \sum_{n=0}^{N-1} \sum_{m=0}^{M-1} X(m,n) \right.$$
$$\left. g([l' - \frac{\tau_p}{T_s} - n(M+L)]T_s) e^{j2\pi m \Delta f([l' - \frac{\tau_p}{T_s} - n(M+L)]T_s)} \right)$$
$$e^{j2\pi(\delta f_c + \nu_p)(l'T_s - \tau_p)}, \tag{6.12}$$

where $l' \in \mathbb{Z}[0 \ \ N(M+L)]$. The n'th OTFS symbol with CP can be collected from the samples of received signal as $r(n'(M+L)+l), \forall l = 0, 1, \cdots, (M+L) - 1$ and can be written as,

$$r(n'(M+L)+l) = \sum_{p=1}^{P} h_p e^{j2\pi\delta f_c \tau_p}$$
$$(\frac{1}{\sqrt{M}} \sum_{n=0}^{N-1} \sum_{m=0}^{M-1} X(m,n) g([(n'-n)(M+L)+l - \frac{\tau_p}{T_s}]T_s)$$
$$e^{j2\pi m \Delta f([(n'-n)(M+L)]T_s)}$$
$$e^{j2\pi m \Delta f([l - \frac{\tau_p}{T_s} - L]T_s)}) e^{j2\pi(\delta f_c + \nu_p)T_s(n'(M+L)+l - \frac{\tau_p}{T_s})} \tag{6.13}$$

It assumed that, after initial coarse synchronization, a residual synchronization error of l_o samples from the starting index of CP removed OTFS symbol as shown in Figure 6.7 exists. It is aslo assumed $(l_0 T_s + \tau_{max}) \leq T_{CP}$, so that no interference is experienced from the neighbouring OTFS symbols. For decoding discrete Fourier transform (DFT) is applied on the CP removed samples of n'th OTFS symbol to obtain the TF data and it is given as,

$$Y(m',n') = \frac{1}{\sqrt{M}} \sum_{k_t=0}^{M-1} r(n'(M+L)+L-l_0+k_t) e^{-j\frac{2\pi m' k_t}{M}}. \tag{6.14}$$

Using (6.13) we can write,

$$Y(m',n') = \frac{1}{M} \sum_{p=1}^{P} h_p e^{j2\pi\delta f_c \tau_p} e^{j2\pi(\nu_p+\delta f_c)T_s(L+n'(M+L)-l_o)}$$

$$e^{-j2\pi(\nu_p+\delta f_c)T_s(\frac{\tau_p}{T_s})} \sum_{n=0}^{N-1}\sum_{m=0}^{M-1} X(m,n)$$

$$e^{-j2\pi \frac{m(l_o+\frac{\tau_p}{T_s}-(n'-n)(M+L))}{M}}$$

$$\sum_{k_t=0}^{M-1} \Big(g([(n'-n)(M+L)+L-l_o+k_t-\frac{\tau_p}{T_s}]T_s)$$

$$e^{j2\pi \frac{k_t(m+(\nu_p+\delta f_c)T_s-m')}{M}} \Big).$$

Since, $g(t)$ is a rectangular pulse,

$$g([(n'-n)(M+L)+L-l_0+k_t-\frac{\tau_p}{T_s}]T_s) =$$

$$\begin{cases} 1, & \text{if } n = n' \ \& \ 0 \le (L-l_0+k_t-\frac{\tau_p}{T_s})T_s \le T \\ 0, & otherwise \end{cases} \tag{6.15}$$

Then, (6.15) can be written as,

$$Y(m',n') = \sum_{p=1}^{P} \tilde{h}_p e^{j2\pi\tilde{\nu}_p T_s(L+n'(M+L)-\frac{\tilde{\tau}_p}{T_s})}$$

$$\sum_{m=0}^{M-1} X(m,n') e^{-j2\pi \frac{m\tilde{\tau}_p}{MT_s}} \Psi((m+\tilde{\nu}_p T_s - m'), M), \tag{6.16}$$

where, $\tilde{h}_p = h_p e^{j2\pi\delta f_c \tau_p}$, $\tilde{\nu}_p = (\nu_p+\delta f_c)$, $\tilde{\tau}_p = (\tau_p+l_0 T_s)$, $\Psi(x,M) \triangleq \frac{1}{M}\sum_{k_t=0}^{M-1} e^{j\frac{2\pi k_t x}{M}}$. When $x \in \mathbb{Z}$, $\Psi(x,M) = \delta([x]_M)$. The TF signal is transformed to De-Do domain using symplectic finite Fourier transform (SFFT) as,

$$y(k',l') = \frac{1}{\sqrt{NM}} \sum_{n'=0}^{N-1}\sum_{m'=0}^{M-1} Y(m',n') e^{-j2\pi[\frac{n'k'}{N}-\frac{m'l'}{M}]}, \tag{6.17}$$

which can be simplified to (as shown in appendix 6.3.3)

$$y(k',l') = \sum_{p=1}^{P} \tilde{h}_p e^{j2\pi\tilde{\nu}_p T_s L} \sum_{k=0}^{N-1}\sum_{l=0}^{M-1} d(k,l)$$

$$\Psi(k-k'+\frac{\tilde{\nu}_p}{\Delta\nu}, N)\Psi(l'-l-\frac{\tilde{\tau}_p}{\Delta\tau}, M) e^{j2\pi \frac{\tilde{\nu}_p(l'-\frac{\tilde{\tau}_p}{\Delta\tau})}{\Delta\nu(M+L)N}}, \tag{6.18}$$

where, $\Delta \nu = \frac{1}{(M+L)NT_s}$ and $\Delta \tau = T_s = \frac{1}{B}$ are Doppler and delay resolution at the receiver. Let $\tilde{\tau}_p = \tilde{l}_p \Delta \tau$ and $\tilde{\nu}_p = \tilde{k}_p \Delta \nu \implies \tilde{\nu}_p T_s = \frac{\tilde{k}_p}{(M+L)N}$, where, $\tilde{l}_p, \tilde{k}_p \in \mathbb{R}$, then (6.18) can be written as,

$$
y(k', l') = \sum_{p=1}^{P} \left\{ \tilde{h}_p e^{j2\pi \frac{\tilde{k}_p L}{(M+L)N}} \sum_{k=0}^{N-1} \sum_{l=0}^{M-1} d(k, l) \right.
$$

$$
\left. \Psi(k - k' + \tilde{k}_p, N) \Psi(l' - l - \tilde{l}_p, M) e^{j2\pi \frac{\tilde{k}_p (l' - \tilde{l}_p)}{(M+L)N}} \right\}
$$

(6.19)

6.2.2 Effects of Residual Synchronization Errors

From (6.18), it may be noted that $y(k', l')$ experiences ISI in both delay and Doppler dimension.

6.2.2.1 Integer Delay and Integer Doppler Values
When $\tilde{l}_p$ and $\tilde{k}_p$ are integers, then (6.18) simplifies as,

$$
y(k', l') = \sum_{p=1}^{P} \left(h_p e^{j2\pi \delta f_c \tau_p} e^{j2\pi \nu'_p T_s L} \right.
$$

$$
\left. d(k' - \tilde{k}_p, l' - \tilde{l}_p) e^{j2\pi \frac{\nu'_p (l' - \frac{\tau'_i}{\Delta \tau})}{\Delta \nu' (M+L)N}} \right)
$$

(6.20)

In AWGN scenario, integer time and frequency errors result in a cyclic shift in delay and Doppler direction respectively, which results in cyclically shifting the origin of De-Do grid to $(l_o, \frac{\delta f_c}{\Delta \nu})$. It can also be observed that, (6.20) resembles the received delay-Doppler signal as described in equation (24) of [101]. Thus, the effect of synchronization error can be considered as a part of the channel itself, with modified channel taps $\tilde{h}_p = h_p e^{j2\pi \delta f_c \tau_p}$, $\tilde{\tau}_p = \tau_p + l_0 T_s$ & $\tilde{\nu}_p = \nu_p + \delta f_c$. Thus, it may be projected estimated channel coefficients may be able to equalize the effects of TVMC and residual synchronization errors.

6.2.2.2 Integer Delay and Fractional Doppler Values
If one observers the the summation terms on the running variables l and k in (6.19) then one can infer the following. When any Doppler or delay value of the modified channel becomes fractional, i.e., $\tilde{k}_p$ or $\tilde{l}_p \notin \mathbb{Z}$, then every symbol would interference in Doppler or delay dimension accordingly. The interference in the Doppler axis is expected to observed almost always since the Doppler values of channel usually do not resolved to integer values. The residual synchronization error, δf_c, can lead to a modified fractional Doppler value even though the actual channel Doppler values (k_p) are integers since $\tilde{k}_p = k_p + \frac{\delta f_c}{\Delta \nu}$. Fractional delay values are not observed since the sampling of the received signal in time domain approximates the effect of fractional channel delay to the nearest integer time bin [101]. Thus, $\tilde{l}_p \in \mathbb{Z} \implies (l' - l - \tilde{l}_p) \in \mathbb{Z} \, \forall \, l' \in [0 \; N-1] \implies \Psi(l - l' + \tilde{l}_p, M) = \delta([l - l' + \tilde{l}_p]_M)$.

Therefore, (6.19) becomes,

$$y(k', l') = \sum_{p=1}^{P} \left\{ \tilde{h}_p e^{j2\pi \frac{\tilde{k}_p L}{(M+L)N}} \sum_{k=0}^{N-1} d(k, [l' - \tilde{l}_p]_M) \right.$$
$$\left. \Psi(k - k' + \tilde{k}_p, N) e^{j2\pi \frac{\tilde{k}_p(l' - \tilde{l}_p)}{(M+L)N}} \right\} \tag{6.21}$$

In (6.21), each received symbol experiences interference from all other symbols in Doppler dimension. In the sections that follow, we describe the construction of the equivalent channel matrix, its estimation and compensation of the effects of synchronization errors described here.

6.2.3 Equivalent Channel Matrix for OTFS Including Synchronization Errors

In this section, we derive the expression of equivalent channel matrix in time domain, which includes the effect of residual synchronization errors and time varying channel. For this, we establish the equivalent system model in matrix-vector form for CP-OTFS. Symbols $d(k, l)$ are arranged in $M \times N$ matrix as,

$$\mathbf{D} = \begin{bmatrix} d(0,0) & d(1,0) & \cdots & d(N-1,0) \\ d(0,1) & d(1,1) & \cdots & d(N-1,1) \\ \vdots & \vdots & \ddots & \vdots \\ d(0,M-1) & d(1,M-1) & \cdots & d(N-1,M-1) \end{bmatrix}. \tag{6.22}$$

De-Do to TF domain conversion (after the ISFFT) is done following $\mathbf{X} = \mathbf{W}_M^\dagger \mathbf{D} \mathbf{W}_N$, where, $\mathbf{X} = \{X(m,n) \mid \forall m \in \mathbb{Z}[0 \ M - 1] \ \& \ n \in \mathbb{Z}[0 \ N - 1]\}$, such that frequency is along m and time is along n. TF domain to time domain signal is obtained using OFDM modulation, as $\mathbf{S} = \mathbf{W}_M \mathbf{X} = \mathbf{D} \mathbf{W}_N$, where, $\mathbf{S} = [\mathbf{s}_0 \ \mathbf{s}_1 \ \cdots \ \mathbf{s}_{N-1}]$ is concatenation of OTFS symbol vectors $\mathbf{s}_i, \forall \ i \in [0 \ N - 1]$. The pulse shaped samples of the signal is written as, $\mathbf{S}_{PS} = \mathbf{G}_T \mathbf{W}_M \mathbf{X}$. $\mathbf{G}_T$ is the pulse shaping matrix[92]. Since $g(t)$ is rectangular, we have $\mathbf{G}_T = \mathbf{I}_M$.

The CP appended signal is given as, $\mathbf{S}_{CP} = \mathbf{B}_{CP} \mathbf{S}_{PS} = \mathbf{B}_{CP} \mathbf{G}_T \mathbf{W}_M \mathbf{X} = \mathbf{B}_{CP} \mathbf{G}_T \mathbf{D} \mathbf{W}_N$, where, $\mathbf{B}_{CP} = \begin{bmatrix} \mathbf{0}_{L \times M - L} & \mathbf{I}_L \\ \mathbf{I}_M & \end{bmatrix}$ is operator for appending CP. Thus, the transmit signal can be given as, $\mathbf{s}_{CP} = vec\{\mathbf{S}_{CP}\} = vec\{\mathbf{B}_{CP} \mathbf{G}_T \mathbf{D} \mathbf{W}_N\}$. Using the identity $vec\{\mathbf{A}_{K \times L} \mathbf{B}_{L \times M}\} = (\mathbf{I}_M \otimes \mathbf{A}) vec\{\mathbf{B}\} = (\mathbf{B} \otimes \mathbf{I}_M) vec\{\mathbf{A}\}$, $\mathbf{s}_{CP} = (\mathbf{I}_N \otimes (\mathbf{B}_{CP} \mathbf{G}_T)) vec\{\mathbf{D} \mathbf{W}_N\} = (\mathbf{I}_N \otimes \mathbf{B}_{CP})(\mathbf{W}_N^T \otimes \mathbf{I}_M) vec\{\mathbf{D}\} = (\mathbf{I}_N \otimes \mathbf{B}_{CP})(\mathbf{W}_N \otimes \mathbf{I}_M) \mathbf{d}$. Therefore,

$$\mathbf{s}_{CP} = \mathbf{A}_{CP} \mathbf{d}, \tag{6.23}$$

where $\mathbf{d} = vec\{\mathbf{D}\}$ is data vector, $\mathbf{A}_{CP} = (\mathbf{I}_N \otimes \mathbf{B}_{CP}) \mathbf{A}$ and $\mathbf{A} = \mathbf{W}_N \otimes \mathbf{I}_M$. We also introduce the transmit signal vector $\mathbf{s} = vec\{\mathbf{S}\}$ without CP added and can be given as

$\mathbf{s} = \mathbf{Ad}$ by putting $\mathbf{B}_{CP} = \mathbf{I}_M$ in (6.23). At the receiver, the noiseless received signal in discrete form [101] can be written as,

$$\mathbf{r}_{CP}(l) = \sum_{p=1}^{P} h_p \mathbf{s}_{CP}(l - l_p) e^{j2\pi \frac{(l-l_p)k_p}{(M+L)N}}, \tag{6.24}$$

where, $l \in \mathbb{Z}[0 \quad ((M + L)N - 1)]$. We collect the samples $\mathbf{r}_{CP}(q(M + L) + l) \forall l \in \mathbb{Z}[0 \quad M + L - 1]$ to obtain qth OTFS symbol vector $\mathbf{r}_{CP}^q$ with CP. Then, the CP removed vector $\mathbf{r}_q$ can be given as,

$$\mathbf{r}_q = \mathbf{R}_{CP}\mathbf{r}_{CP}^q, \text{ where } \mathbf{R}_{CP} = \begin{bmatrix} \mathbf{0}_{M \times L} & \mathbf{I}_M \end{bmatrix}. \tag{6.25}$$

Therefore,

$$\mathbf{r}_q(l) = \mathbf{r}_{CP}(q(M + L) + L + l), \ l \in [0 \quad M - 1]. \tag{6.26}$$

With the introduction of CFO at the receiver, the samples of qth received OTFS symbol $\mathbf{r}_q^f(l)$ is

$$\mathbf{r}_q^f(l) = \mathbf{r}_{CP}(q(M + L) + L + l)e^{j2\pi \frac{k_0(q(M+L)+L+l)}{N(M+L)}}, \tag{6.27}$$

where, $k_0 = \frac{\delta f_c}{\Delta\nu}$. With residual time synchronization error of l_o samples,

$$\tilde{\mathbf{r}}_q(l) = \mathbf{r}_q^f(l - l_o)$$

From (6.24), (6.26) and (6.27),

$$\tilde{\mathbf{r}}_q(l) = e^{j2\pi \frac{k_0(q(M+L)+L+l-l_o)}{N(M+L)}} \sum_{p=1}^{P} h_p e^{j2\pi \frac{k_p q}{N}} \\ e^{j2\pi \frac{(L+l-l_o-l_p)k_p}{(M+L)N}} \mathbf{s}_{CP}(q(M + L) + L + l - l_o - l_p). \tag{6.28}$$

We assume $l_o + l_\tau < L$. Therefore,

$$\tilde{\mathbf{r}}_q(l) = \sum_{p=1}^{P} h_p e^{j2\pi \frac{l_p k_0}{(M+L)N}} e^{j2\pi \frac{(k_p+k_0)q}{N}} e^{j2\pi \frac{(L+l-l_o-l_p)(k_p+k_0)}{(M+L)N}} \\ \mathbf{s}_q([l - l_p - l_o]_M). \tag{6.29}$$

Then,

$$\tilde{\mathbf{r}}_q = \begin{bmatrix} \tilde{\mathbf{r}}_q(0) \\ \tilde{\mathbf{r}}_q(1) \\ \vdots \\ \tilde{\mathbf{r}}_q(M - 1) \end{bmatrix} \\ = \sum_{p=1}^{P} h_p e^{j2\pi \frac{l_p k_0}{(M+L)N}} \Pi^{l_p+l_o} \Delta^{k_p+k_0}$$

$$e^{j2\pi \frac{(k_p+k_0)(L-l_p-l_o)}{(M+L)N}} e^{j2\pi \frac{k_p+k_0}{N}q} \mathbf{s}_q, \tag{6.30}$$

where, $\Pi = circ\{[0 \ 1 \ 0 \cdots 0]_{M\times 1}^{\mathrm{T}}\}$ is a circulant delay matrix and $\Delta = \mathrm{diag}\{[1 \ e^{j2\pi \frac{1}{(M+L)N}} \ \cdots e^{j2\pi \frac{M-1}{(M+L)N}}]^{\mathrm{T}}\}$ is a diagonal Doppler matrix. Let, $\tilde{l}_p = l_p + l_o$, $\tilde{k}_p = k_p + k_0$ and $\tilde{h}_p = h_p e^{j2\pi \frac{l_p k_0}{(M+L)N}}$. When we include noise,

$$\tilde{\mathbf{r}}_q = \tilde{\mathbf{H}}_q \mathbf{s}_q + \mathbf{v}_q, \tag{6.31}$$

where, $\mathbf{v}_q$ is M length Gaussian noise vector with elemental variance σ_v^2 and

$$\tilde{\mathbf{H}}_q = \sum_{p=1}^{P} \tilde{h}_p \Pi^{\tilde{l}_p} \Delta^{\tilde{k}_p} e^{j2\pi \frac{\tilde{k}_p(L-\tilde{l}_p)}{(M+L)N}} e^{j2\pi \frac{\tilde{k}_p}{N}q}. \tag{6.32}$$

The above can be modified as,

$$\tilde{\mathbf{H}}_q = \sum_{l=0}^{\tilde{l}_\tau} \Pi^l \sum_{k \in \tilde{k}_{\nu_l}} \mathrm{diag}\{\tilde{h}_{l,k} e^{j2\pi \frac{k(L-l)}{(M+L)N}} e^{j2\pi \frac{k}{N}q} [1 \ e^{j2\pi \frac{k}{(M+L)N}} \ \cdots e^{j2\pi \frac{(M-1)k}{(M+L)N}}]^{\mathrm{T}}\}, \tag{6.33}$$

where, $\tilde{l}_\tau = l_\tau + l_0$ is maximum excess delay bin value of channel. We let k_{ν_l} be the set of Doppler indices for the lth channel tap such that $P = \sum_{l=0}^{\tilde{l}_\tau} k_{\nu_l}$. Then, $\tilde{k}_{\nu_l} = \{(k+k_0)| \ k \in k_{\nu_l}\}$ and

$$\tilde{h}_{l,k} = \begin{cases} \tilde{h}_p, \text{if } \tilde{l}_p = l \text{ and } \tilde{k}_p = k \\ 0, \text{otherwise.} \end{cases}$$

Therefore, the concatenation of CP removed vectors can be given as,

$$\tilde{\mathbf{r}} = \begin{bmatrix} \tilde{\mathbf{r}}_0 \\ \tilde{\mathbf{r}}_1 \\ \vdots \\ \tilde{\mathbf{r}}_{N-1} \end{bmatrix} = \underbrace{\begin{bmatrix} \tilde{\mathbf{H}}_0 & & & \\ & \tilde{\mathbf{H}}_1 & & \\ & & \ddots & \\ & & & \tilde{\mathbf{H}}_{N-1} \end{bmatrix}}_{\mathbf{H}} \underbrace{\begin{bmatrix} \mathbf{s}_0 \\ \mathbf{s}_1 \\ \vdots \\ \mathbf{s}_{N-1} \end{bmatrix}}_{\mathbf{s}} + \mathbf{v}, \tag{6.34}$$

which can be written as, $\tilde{\mathbf{r}} = \mathbf{Hs} + \mathbf{v} = \mathbf{HAd} + \mathbf{v}$, $\mathbf{v}$ being MN length concatenated white Gaussian noise vector. Equation (6.31) suggests that the effect of synchronization errors can be considered as part of time domain channel matrix. Equation (6.32) shows that the structure of the channel matrix is invariant to the introduction of synchronization errors which is an added advantage as the number of elements in the matrix does not change with residual synchronization errors, and hence, the sparsity of the matrix is unaltered. Next, an algorithm to estimate this equivalent channel matrix is described, however, first a short justification for choosing time domain channel estimation over De-Do domain processing is given.

6.2.3.1 OTFS Channel Matrices

De-Do channel matrix $\mathbf{H}_{DD}$ and time domain channel matrix $\mathbf{H}$ are related as [92], $\mathbf{H}_{DD} = \mathbf{A}^\dagger \mathbf{H} \mathbf{A}$. Using the definition of $\mathbf{A} = \mathbf{W}_N \otimes \mathbf{I}_M$, $\mathbf{H}_{DD}$ can be simplified as a block matrix with blocks of size $M \times M$ and can be written as, $\mathbf{H}_{DD} =$

$$\begin{bmatrix} \mathbf{H}_{dd}^{0,0} & \cdots & \mathbf{H}_{dd}^{(N-1),0} \\ \vdots & \ddots & \vdots \\ \mathbf{H}_{dd}^{0,(N-1)} & \cdots & \mathbf{H}_{dd}^{(N-1),(N-1)} \end{bmatrix},$$ whose $(l,k)^{\text{th}}$ block can be given as, $\mathbf{H}_{dd}^{l,k} =$

$\sum_{q=0}^{N-1} \bar{w}_{q,l} w_{q,k} \tilde{\mathbf{H}}_q$, where, $w_{l,k} = e^{j2\pi \frac{lk}{N}}$.

Using $\tilde{\mathbf{H}}_q$ as in (6.33), $\mathbf{H}_{dd}^{l,k}$ can be further simplified as,

$$\mathbf{H}_{dd}^{l,k} = \begin{cases} \sum_{p=1}^{P} \tilde{h}_p \Pi^{\tilde{l}_p} \Delta^{\tilde{k}_p} e^{j2\pi \frac{\tilde{k}_p(L-\tilde{l}_p)}{(M+L)N}} \delta([k-l+\tilde{k}_p]_N), \\ \text{for } \tilde{k}_p \in \mathbb{Z} \\[2mm] \sum_{p=1}^{P} \tilde{h}_p \Pi^{\tilde{l}_p} \Delta^{\tilde{k}_p} e^{j2\pi \frac{\tilde{k}_p(L-\tilde{l}_p)}{(M+L)N}} \Psi(k-l+\tilde{k}_p, N), \\ \text{for } \tilde{k}_p \notin \mathbb{Z} \end{cases}$$

We can infer the following from the above equation. When channel Doppler values are resolved in integer Doppler bins, each row of $\mathbf{H}_{DD}$ contains P matrices where $l - k = \tilde{k}_p$, $\forall p = 1, \cdots, P$ which results in NPN_e number of nonzero elements in $\mathbf{H}_{DD}$. However, when fractional Doppler is observed, the De-Do channel matrix $\mathbf{H}_{DD}$ contains $N^2 N_e$ number of elements. Thus, structure of channel matrix varies with the nature of channel Doppler values which is not observed in time domain channel matrix. Therefore, $\mathbf{H}$ is at least P times and at max N times sparse than $\mathbf{H}_{DD}$. Therefore, equalization with $\mathbf{H}$ will result in lower complexity than equalization with $\mathbf{H}_{DD}$.

6.2.4 Estimation of Equivalent Channel Matrix

From (6.33), we can write,

$$\tilde{\mathbf{H}}_q = \sum_{l=0}^{\tilde{l}_\tau} \Pi^l \mathbf{H}_{q,l}, \text{ where,} \tag{6.35}$$

$$\mathbf{H}_{q,l} = diag\{[h(q(M+L)+L,l)\ h(q(M+L)+L+1,l) \\ \cdots\ h(q(M+L)+M+L-1,l)]^{\text{T}}\}, \tag{6.36}$$

where,

$$h(n,m) = \sum_{k \in \tilde{k}_{\nu_m}} h_{m,k} e^{j2\pi \frac{k_0 m}{N(M+L)}} e^{j2\pi \frac{k(n-m)}{N(M+L)}}, \tag{6.37}$$

$\forall\, n \in [0\ N(M+L)-1], m \in [0\ \tilde{l}_\tau]$.

6.2.4.1 Pilot Structure in Delay-Doppler Domain

The pilot structure described for OTFS in [98] is extended for use here. The pilot is a 2-dimensional (2D) impulse in De-Do domain i.e., $d(k,l) = \sqrt{P_{PLT}}\delta(k - K_p, l - L_p)$, $,\forall k \in [0 \ N-1], l \in [L_p - L \ L_p + L - 1]$, where $P_{PLT} = N_p = 2NL$ is the pilot power. At the receiver, the received De-Do signal corresponding to pilot signal, with synchronization error is,

$$y(k',l') = \sum_{p=1}^{P} \left\{ \tilde{h}_p e^{j2\pi \frac{\tilde{k}_p L}{(M+L)N}} \sum_{k=0}^{N-1} \sqrt{P_{PLT}}\delta(k - K_p, \right.$$

$$\left. [l' - L_p - \tilde{l}_p]_M)\Psi(k - k' + \tilde{k}_p, N)e^{j2\pi \frac{\tilde{k}_p(l' - \tilde{l}_p)}{(M+L)N}} \right\} \tag{6.38}$$

From the above, one may infer that the transmitted 2D impulse pilot, after going through the channel, spreads over the entire Doppler axis while the spread in delay is limited to $(L_p + \tilde{l}_\tau)$ starting from L_p. The received signal samples $y(k',l')$, $\forall k \in [0 \ N-1], l \in [L_p \ L_p + L - 1]$ are collected for channel estimation. This part of the received signal contains the response of the channel to the 2D-De-Do impulse pilot signal, which is not interfered by data symbols. This is because it has been assumed that $L \geq l_\tau + l_o - 1$.

6.2.4.2 Channel Estimation

Some necessary theorems are given below , which are required to obtain the time domain channel estimates from the received time domain signal. Theorem 6.1 relates the time domain channel coefficients and the received De-Do signal $y(k',l')$. Theorem 6.2 establishes the relationship between the estimates of time domain channel coefficients and the received time domain signal $\mathbf{r}_{CP}(l)$.

Theorem 6.1 *The intermittent time-domain channel coefficients are directly proportional to the N point IDFT values obtained from pilot section of received De-Do grid, i.e.,* $h(\alpha(M + L) + L + L_p + l, l) = e^{-j2\pi \frac{\alpha K_p}{N}}\left(\sum_{k'=0}^{N-1} y(k', L_p + l)e^{j2\pi \frac{\alpha k'}{N}} \right), \forall \ \alpha \in \mathbb{Z}[0 \ N-1], \ l \in \mathbb{Z}[0 \ \tilde{l}_\tau - 1].$

Proof is given in section 6.3.1.

Theorem 6.2 *The intermittent time-domain channel coefficients are directly proportional to the samples of received signal, i.e,*
$\hat{h}(\alpha(M + L) + L_p + l, l) = e^{-j2\pi \frac{\alpha K_p}{N}}\mathbf{r}_{CP}(\alpha(M + L) + L + L_p + l), \ \alpha \in [0 \ N-1]$

Proof is given in section 6.3.2.

From the above theorem, one can see that the channel coefficients at time instances $(\alpha(M + L) + L_p + l)T_s$, $\alpha \in [0 \ N-1]$ for lth channel tap can be directly estimated from $\mathbf{R}(l,k) = \mathbf{r}_k(l)$, as in Algorithm 5, without going through the FFT-ISFFT path to reach De-Do domain as described in [98]. This creates the opportunity for estimating

---

Algorithm 5 Esitmation of $\tilde{\mathbf{H}}_q$

1:   Given : The received signal $r(l)$

2:   Output : $\hat{\tilde{\mathbf{H}}}_q$

3:   $\mathbf{R}(l,k) = \mathbf{r}_k(l)$

4:   **for** $l' = 0 : L$ **do**

5:     $\gamma = 0$

6:     **for** $k = 0 : N - 1$ **do**

7:       $\gamma = \gamma + \|\mathbf{R}(L_p + l', k)\|^2$

8:     **end for**

9:     **if** $\gamma > 3\sigma_n^2$ **then**

10:       $l = [l\ \ l']$

11:       $\hat{h}(n(M + L) + L + L_p + l', l') = \frac{1}{\sqrt{P_{PLT}}}\mathbf{R}(L_p + l', n)$

12:       $\hat{h}(n, l') = \text{spline\_interpolate}(\{\hat{h}(n(M + L) + L + L_p + l', l') \mid n \in [0\ \ N - 1]\}, [L + L_p + l' : M + L : N(M + L)], [0 : N(M + L) - 1])$

13:     **end if**

14:   **end for**

15:   **for** $q = 0 : N - 1$ **do**

16:     $\hat{\mathbf{H}}_q = \mathbf{0}_{M \times M}$

17:     **for** $i \in l$ **do**

18:       $\hat{\mathbf{H}}_{q,i} = diag\{\{\hat{h}(q(M + L) + L + l', i) | l' \in [0\ \ M - 1]\}\}$

19:       $\hat{\tilde{\mathbf{H}}}_q = \hat{\tilde{\mathbf{H}}}_q + \Pi^i \hat{\mathbf{H}}_{t,\tau}^{q,i}$

20:     **end for**

21: **end for**

---

time domain channel coefficients from De-Do domain embedded pilot. Using the estimated channel coefficients one can obtain the channel tap values at time instances nT_s, $n \in \mathbb{Z}[0\ \ N(M + L) - 1]$ through interpolation. The coefficients to estimate (6.36) resembles a sum of sinusoids. Thus, polynomial interpolation techniques (spline [117]) are used. In other words,

$$
\begin{aligned}
h(n, l) \ = \text{spline\_interpolate}\Big([h(L_p + l, l) \\
h(M + L + L_p + l, l) \\
\cdots h((N - 1)(M + L) + L_p + l, l)]^T, \\
\cdots [0\ M + L(N - 1)(M + L)]^T \\
, [0\ 1\ 2\ \cdots\ (N)(M + L) - 1]^T\Big),
\end{aligned}
\tag{6.39}
$$

$\forall\, l \in [0\ l_\tau - 1]$, where spline $-$ interpolate (using 'interpl1' inbuilt function in Matlab®), returns the interpolated signal at points u when y is the part of signal known at points x. Algorithm 5 describes how channel estimates for the entire time duration are generated from the received signal as well as how to estimate qth channel matrices $\hat{\mathbf{H}}_{q,i}$ & $\hat{\tilde{\mathbf{H}}}_q$, which are described in (6.36) & (6.35), respectively.

6.2.4.3 Time Domain Interpretation of the Channel Estimation

The frame structure used in De-Do domain yields an impulse train in the time domain modulated by sinusoid, which does not interfere with the equivalent time domain data as shown in Figure 6.8. It is given as,

$$s(n) = s_p(n) + s_d(n). \tag{6.40}$$

Symbols $s_d(n)$ and $s_p(n)$ are the corresponding time domain equivalent signal of the the De-Do data and pilot. Since, pilot is a 2D discrete impulse at location (K_p, L_p),

$$s_p(n) = \sqrt{\frac{P_{PLT}}{N}} e^{j2\pi \frac{K_p n}{N(M+L)}} \sum_{\alpha=0}^{N-1} \delta(n - (\alpha(M+L) + L + L_p)) \tag{6.41}$$

Using (6.24), (6.37), (6.40) and (6.41), the received signal is,

$$r(n) = \sum_{l=0}^{l_\tau - 1} h(n,l) \left(\sqrt{\frac{P_{PLT}}{N}} e^{j2\pi \frac{K_p(n-l)}{N(M+L)}} \sum_{\alpha=0}^{N-1} \right.$$

$$\delta(n - l - (\alpha(M+L) + L + L_p))) + \sum_{l=0}^{l_\tau - 1} h(n,l) s_d(n - l)$$

which can be simplified as,

$$r(n) = \sqrt{\frac{P_{PLT}}{N}} \sum_{\alpha=0}^{N-1} h(n, n - (\alpha(M+L) + L + L_p)) \times$$

$$e^{j2\pi \frac{K_p(\alpha(M+L) + L + L_p)}{N(M+L)}} + \sum_{l=0}^{l_\tau - 1} h(n,l) s_d(n - l).$$

With length of CP being greater than τ_{max},

$$r(\alpha(M+L) + L + L_p + l) = \sqrt{\frac{P_{PLT}}{N}} \times$$

$$h(\alpha(M+L) + L_p + l, l) e^{j2\pi \frac{K_p(\alpha(M+L) + L + L_p)}{N(M+L)}},$$

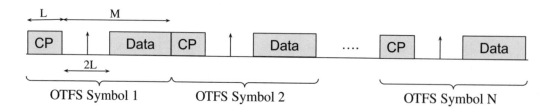

Figure 6.8 Time domain CP-OTFS frame.

which implies the result as obtained using theorem 6.2. From this time domain interpretation one can consider the time domain frame as pilots embedded in each symbol. After such a pilot passes through a channel, one can obtain the channel impulse response directly. This can be interpolated to obtain the channel coefficients at all time instants as described earlier.

6.2.5 LMMSE Equalization

The LMMSE equalization is as described in Section 4.2.2. The LMMSE equalization of $\mathbf{r}$ in (6.34) results in estimated data vector $\hat{\mathbf{d}}$ as given in 3.23. When $g(t)$ is rectangular, $\mathbf{A}$ becomes unitary. Thus, (3.23) becomes (4.17). It is mentioned in Section 4.2.2 that it is desired to reduce the complexity of $\mathbf{r}_{ce} = \mathbf{H}_{eq}\mathbf{r}$.

It is evident from (6.34), that $\mathbf{H}$ matrix is a block diagonal matrix with blocks $\tilde{\mathbf{H}}_q$ of size $M \times M$. This leads to $\Psi = \mathbf{H}\mathbf{H}^\dagger + \frac{\sigma_\nu^2}{\sigma_d^2}\mathbf{I} = diag\{\Psi_0, \ \Psi_1, \ \cdots \ \Psi_{N-1}\}$, which is a block diagonal matrix with blocks Ψ_q of size $M \times M$. It is well known that the inverse of a block diagonal matrix is also a block diagonal matrix. In addition to that, the inverse of a block diagonal matrix can be computed using the inverse of individual blocks. Similar to the decomposition of $\mathbf{r}$ into $\mathbf{r}_q(\text{s})$, $\mathbf{r}_{ce}$ can also be written as $\mathbf{r}_{ce} = [\mathbf{r}_{ce,0}^{\mathrm{T}} \ \mathbf{r}_{ce,1}^{\mathrm{T}} \cdots \mathbf{r}_{ce,N-1}^{\mathrm{T}}]^{\mathrm{T}}$, where $\mathbf{r}_{ce,q} = [\mathbf{r}_{ce}(q(M+L)+L) \ \mathbf{r}_{ce}(q(M+L)+L+1) \cdots \mathbf{r}_{ce}(q(M+L)+(M+L)-1)]^{\mathrm{T}}$ is the qth, $q \in [0 \ N-1]$ channel equalized vector. Thus we can write,

$$\mathbf{r}_{ce,q} = \tilde{\mathbf{H}}_q^\dagger[\tilde{\mathbf{H}}_q\tilde{\mathbf{H}}_q^\dagger + \frac{\sigma_\nu^2}{\sigma_d^2}\mathbf{I}]^{-1}\mathbf{r}_q, \ q \in \mathbb{Z}[0 \ N-1], \tag{6.42}$$

which can be computed using inversion and multiplication of $M \times M$ matrices. The required complexity is of $O(NM^3)$. Generally, the value of M is in the order of 100's. Although the above simplifications significantly reduce complexity, yet LMMSE processing remains a computational burden. The structure of Ψ_q involved in channel equalization described earlier is explored in order to further reduce the complexity.

6.2.5.1 Structure of $\Psi_q = [\tilde{\mathbf{H}}_q\tilde{\mathbf{H}}_q^\dagger + \frac{\sigma_\nu^2}{\sigma_d^2}\mathbf{I}]$

Using (6.32), $\tilde{\mathbf{H}}_q\tilde{\mathbf{H}}_q^\dagger$ can be expressed as,

$$\tilde{\mathbf{H}}_q\tilde{\mathbf{H}}_q^\dagger = \left(\sum_{p=1}^{P} \tilde{h}_p e^{j2\pi\frac{\tilde{k}_p(L-\tilde{l}_p)}{(M+L)N}}\Delta^{\tilde{k}_p}\Pi^{\tilde{l}_p}e^{j2\pi\frac{\tilde{k}_p}{N}q}\right)$$

$$\left(\sum_{s=1}^{P} \bar{\tilde{h}}_s e^{-j2\pi\frac{\tilde{k}_s(L-\tilde{l}_s)}{(M+L)N}}\Pi^{-\tilde{l}_s}\Delta^{-\tilde{k}_s}e^{j2\pi\frac{\tilde{k}_s}{N}(-q)}\right) \tag{6.43}$$

Since Π is a circulant matrix, therefore $\Pi^{l_p} = \mathbf{W}\Delta^{-l_p}\mathbf{W}^\dagger$. Thus, $\tilde{\mathbf{H}}_q\tilde{\mathbf{H}}_q^\dagger = \sum_{\substack{p=1 \\ p=s}}^{P}|\tilde{h}_p|^2\mathbf{I} +$

$\sum_{p=1}^{P}\sum_{\substack{s=1 \\ p\neq s}}^{P}c_{p,s}\Pi^{l_p-l_s}\Delta^{k_p-k_s}e^{j2\pi q\frac{k_p-k_s}{N}}$, where $c_{p,s} = \tilde{h}_p\bar{\tilde{h}}_s e^{j2\pi\frac{-\tilde{k}_p\tilde{l}_p+\tilde{k}_s\tilde{l}_s}{(M+L)N}}e^{j2\pi\frac{L(\tilde{k}_p-\tilde{k}_s)}{(M+L)N}}$.

Hence, we can write

$$\Psi_q = \sum_{\substack{p=1 \\ p=s}}^{P}(|\tilde{h}_p|^2 + \frac{\sigma_\nu^2}{\sigma_d^2})\mathbf{I} + \sum_{\substack{p=1 \\ p\neq s}}^{P}\sum_{s=1}^{P}c_{p,s}\Pi^{\tilde{l}_p - \tilde{l}_s}\Delta^{\tilde{k}_p - \tilde{k}_s}e^{j2\pi q\frac{\tilde{k}_p - \tilde{k}_s}{N}}. \tag{6.44}$$

From the above, it can be concluded that the maximum shift of diagonal elements in Δ can be $\pm(\tilde{l}_\tau - 1)$. It can also be noted that due to the cyclic nature of the shift, Ψ_q is quasi-banded with bandwidth of $2\tilde{l}_\tau - 1$. As $\tilde{l}_\tau \ll M$, Ψ_q is also sparse for typical wireless channel. Structure of Ψ_q is similar to the channel matrix of RCP OTFS as described in equation (13) specified in [94]. Thus, Ψ_q^{-1} can be computed using LU factorization of Ψ_q in a similar way as described in Sec. III B of [94], i.e. $\Psi_q = \mathbf{L}_q\mathbf{U}_q$.

6.2.5.2 Computation of $\hat{\mathrm{d}}$
Using LU decomposition of Ψ_q, $\mathbf{r}_{ce,q}$ is simplified to,

$$\mathbf{r}_{ce,q} = \tilde{\mathbf{H}}_q^\dagger \overbrace{\mathbf{U}_q^{-1}\underbrace{\mathbf{L}_q^{-1}\mathbf{r}_q}_{\mathbf{r}_q^{(1)}}}^{\mathbf{r}_q^{(2)}}.$$

As $\mathbf{L}_q$ is a quasi-banded lower triangular matrix, $\mathbf{r}_q^{(1)} = \mathbf{L}_q^{-1}\mathbf{r}_q$ can be computed using low complexity forward substitution as explained in Algorithm 2 in SEction 4.2. Algorithm 3 of Section 4.2 can be used to evalaute $\mathbf{r}_q^{(2)} = \mathbf{U}_q^{-1}\mathbf{r}_q^{(1)}$. Using the definition of $\mathbf{H}_q$, $\mathbf{r}_{ce,q} = \tilde{\mathbf{H}}_q^\dagger\mathbf{r}_q^{(2)}$ can be written as,

$$\mathbf{r}_{ce,q} = \sum_{p=1}^{P}\bar{h}_p\Delta^{-k_p}\underbrace{\Pi^{-l_p}\mathbf{r}_q^{(2)}}_{\text{circular shift}}.$$

To compute $\mathbf{r}_{ce,q}$, $\mathbf{r}_q^{(2)}$ is first circularly shifted by '$-l_p$' and then multiplied by $\bar{h}_p\text{diag}\{\Delta^{-k_p}\}$ using point-to-point multiplication for each path p. All vectors obtained above are summed to obtain $\mathbf{r}_{ce,q}$. Then, $\{\mathbf{r}_{ce,q}\}_{q=0}^{N-1}$ are concatenated to obtain $\mathbf{r}_{ce}$. Finally, $\hat{\mathbf{d}} = \mathbf{A}^\dagger\mathbf{r}_{ce}$ can be implemented using M number of N-point FFTs (Sec. III-C, [94]).

6.2.5.3 Computation Complexity
The number of CMs required to implement the LMMSE algorithm is $\frac{MN}{2}\log_2 N + MN[2l_\tau^2 + 2P^2k_\nu + 9l_\tau - Pk_\nu - 3] + N[\frac{2}{3}l_\tau^3 + 2l_\tau + P]$. The order of complexity achieved through the above receiver design is $MN\log(MN)$, which is significantly lower than the direct implementation, which is of the order of M^3N^3. Our proposed receiver requires around 10^7x lower CMs than the direct implementation following (3.23), if we consider a typical OTFS system with $\Delta f = 15\ KHz$, $f_c = 4\ GHz$, $N = 128$, $M = 512$, speed of 500 Kmph and the extended vehicular A (EVA) 3GPP channel model [30] with $P = 9$ and $\tau_{max} = 2.51\ \mu\ sec$.

6.2.6 LDPC Coded LMMSE-SIC Reciever

The successive interference cancellation receiver for LDPC coded OTFS, which was presented in Section 4.3.2, now uses the estimated channel coefficients instead of ideal channel coefficients.

6.2.7 Unified Framework for Orthogonal Multicarrier Systems

It was shown in Chapter 3 that OTFS can also be viewed as block OFDM. It is also known by now that both OFDM and OTFS are orthogonal waveforms. In Section 4.2.2.4 it was shown how the LMMSE receiver can be used for OFDM as well. The use of LMMSE receiver for GFDM, as well as for OFDM was described in Section 5.5.1. In this section, a generalized framework for cyclic prefixed orthogonal waveforms is further developed. This helps in having a unified time domain frame for orthogonal waveforms.

This framework will be able to help different waveforms take advantage of the channel *estimation and equalization algorithms described earlier in this chapter. Let* $\mathbf{D}$ *be the data* matrix of size $(M) \times N$. Then, the transmit signal which can be of the form

$$\mathbf{S}_d = \{\mathfrak{B}\mathbf{D}\mathfrak{C}\} = \mathbf{A}vec\{\mathbf{D}\}, \tag{6.45}$$

where, $\mathfrak{B}$ and $\mathfrak{C}$ are modulation specific matrices. $\mathbf{A}$ in (3.23) can be computed using $\mathbf{A} = (\mathfrak{C} \otimes \mathbf{I})(\mathbf{I} \otimes \mathfrak{B})$. We define $\mathbf{S}_p$ as the constant pilot matrix of size $2L \times N$ as,

$$\mathbf{S}_p = \begin{bmatrix} \mathbf{0}_{L \times N} \\ [1\ 1\ \cdots\ 1]_{1 \times N} \\ \mathbf{0}_{L-1 \times N} \end{bmatrix}. \tag{6.46}$$

The combined time domain matrix with N symbols each containing M samples can be given as, $\mathbf{S} = \begin{bmatrix} \mathbf{S}_p \\ \mathbf{S}_d \end{bmatrix}$. We append CP as described before, i.e., $\mathbf{S}_{CP} = \mathbf{B}_{CP}\mathbf{S}$. Then, the transmit vector can be written as $\mathbf{s} = vec\{\mathbf{S}_{CP}\}$. This transmit vector will resemble the time domain frame illustrated in Figure 6.8. Therefore, the same time domain channel estimation and low complexity LMMSE equalization detailed above can be used to equalize the combined effects of channel and residual synchronization errors.

CP-OTFS considered in this chapter fits in this framework when one sets $\mathfrak{B} = \mathbf{I}_M$ and $\mathfrak{C} = \mathbf{W}_N$ in above equations. Similarly, when $\mathfrak{B} = \mathbf{W}_M$ and $\mathfrak{C} = \mathbf{I}_N$, the above system becomes an OFDM system with no difference in the equalization techniques at the receiver. Hence, this generalized description of system paves the way to realize a flexible communication system which can be used to change the waveform with changing nature of channel, for e.g., the transceiver pair can use OFDM in low mobility scenarios while switching to OTFS under high speed scenario.

6.2.8 Results

Now, we analyze the performance of LDPC coded CP-OTFS system with time domain channel estimation and equalization in presence of residual FTO and CFO errors. The

Table 6.2   Simulation parameters.

Parameter	Value
Carrier frequency(f_c)	6 GHz
Bandwidth(B)	7.68 MHz
Frame time(T_f)	8.7 ms
Subcarrier bandwidth(Δf)	15 KHz
OTFS parameters	M=512, N=128 , $\Delta \nu = 117.18$ Hz, $\Delta \tau = 130.21$ ns
Pilot location	K_p=0, L_p=10
Channel model	EVA [30]
CP duration	T_{CP} =1.302 μs, $L = 10$
UE speed	500 Kmph
Modulation	16-QAM
FEC (LDPC)	Coderate $\frac{2}{3}$, codeword length 648

performance of an equivalent OFDM system is also made since a unified framework is now available. The use of the developed time domain channel estimation when applied to RCP-OFTS is also verified, however, without residual synchronization errors. The simulation parameters are given in Table 6.2.

The legend "Ideal" marks curves which show the performance of LMMSE equalization with ideal channel estimate without any residual synchronization error. The legend "MMSE" indicates use of LMMSE equalizer which uses estimated channel coefficeints. The label "sic" indicates use of SIC receiver with estimated channel coefficients. The numeral following these key words indicates the number of Doppler taps per delay tap ("Dpt") used in evaluation. The legend "synch" denotes evaluation cases, where $l_0 = 2$ and $k_0 = 20$. In case of "synch", estimated channel coefficients are used in LMMSE equalization and "sic". It may be noted that normalized CFO error $k_0 = \frac{\delta f_c}{\Delta \nu} = 20$ results in $\delta f_c = k_0 \Delta \nu = 2.33$ KHz.

6.2.8.1 Block Error Rate (BLER) Performance

The BLER performance of CP-OTFS system is shown in Figure 6.9. For "Ideal", it is observed that as "Dpt" increases, the performance of the system improves. The improvement in BLER performance with increasing "Dpt" can be attributed to Doppler diversity. With increasing "Dpt" there is increase in the number of independent channel paths. It also indicates that the LMMSE receiver is able to extract this diversity from the received signal.

Next, we consider the performance of CP-OTFS with the time domain channel estimation algorithm but without residual synchronization error. The degradation in the performance of "MMSE" with respect to "Ideal" is limited to ≈ 1 dB for all Dpt considered. The performance of "sic" seen to be very close to that of "Ideal".

It can be seen that "MMSE-synch", for 1 Dpt , suffers by nearly 0.5 dB, when compared to no synchronization errors, which is about 1.5 dB for 3 Dpt. In case of "sic" it can be observed that the performance is not much deviated from "Ideal" untill error floor starts to appear at higher SNRs. It may also be stated that CP-OTFS is fully immune to residual synchronization error within the performance limits of required reliability.

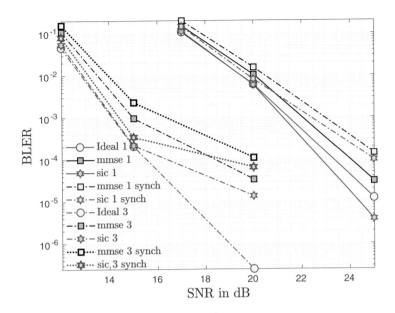

Figure 6.9 BLER Vs SNR (dB) for CP-OTFS with 16-QAM, ldpc code word length 648, code rate=$\frac{2}{3}$ at 500 Kmph.

The BLER vs SNR performance of CP-OTFS for 64-QAM is exposed in Figure 6.10. It can be seen that at a BLER of 10^{-1}, about 7 dB additional SNR required over 16-QAM. At BLER of 10^{-3} and below it is around 10 dB.

The significant gain the SIC has over single stage MMSE receiver is predominantly observable for higher order QAM.

Thus, it can observed that SIC receiver works effectively with the above time domain channel estimation technique and provides support for very high data rate CP-OTFS systems.

Since RCP-OTFS is spectrally more efficient than CP-OTFS, its performance with the time domain channel estimation algorithm is presented in Figure 6.11. For RCP-OTFS, the CP length is zero for all except the first CP, which is drawn from the entire OTFS block. Effects of residual synchronization errors are not included since RCP-OTFS with residual synchronization errors require further investigation.

It can be seen that upto a BLER of 10^{-2} there is not much difference in performance between CP-OTFS and RCP-OTFS. Although RCP-OTFS is expected to encounter a large amount of interference owing to lack of CP between the OFDM symbols, however, the pilot structure, explained above, helps to reduce inter-OFDM-symbol interference. At higher SNR, RCP-OTFS is found to perform slightly better than CP-OTFS. This can be ascribed to improved channel estimation owing to lower pilot sampling interval.

Now, lets take a look at the performance of OFDM system, which is described in Section 6.2.7, as given in Figure 6.12. It may not easily noted that SIC does not provide any notable improvement. The reason being that in OFDM systems, the QAM symbols

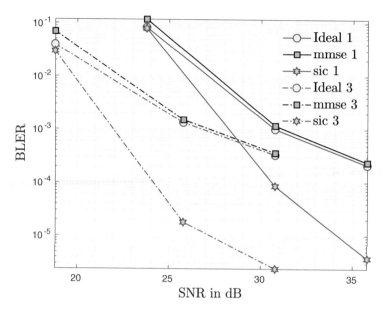

Figure 6.10   BLER Vs SNR for 64-QAM with ldpc code word length 648, code rate=$\frac{2}{3}$ at 500 Kmph.

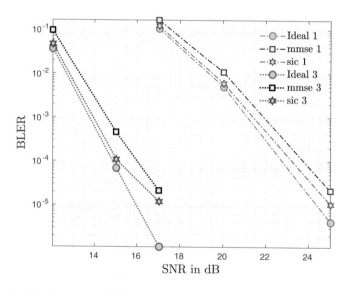

Figure 6.11   BLER Vs SNR (dB) for RCP-OTFS with 16-QAM, ldpc code word length 648, code rate=$\frac{2}{3}$ at 500 Kmph.

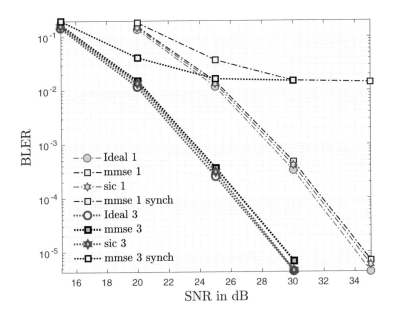

Figure 6.12 BLER Vs SNR for CP-OFDM with 16-QAM, ldpc code word length 648, code rate=$\frac{2}{3}$ at 500 Kmph.

are carried on each subcarrier whereas in OTFS they are spread over the entire TF space. Another interesting observation is that with Dpt of 3, the performance is better by nearly 5 dB over the scenario of 1 Dpt.

When one compares OFDM against CP-OTFS, one may first consider the "Ideal", "MMSE" and "sic" cases without residual synchronization errors.

It can be seen that for Dpt = 1, OFDM looses by nearly 3 dB at higher BLER. At BLER of 10^{-2}, the gap increases to nearly 5 dB, thereafter, continutes to increase further as SNR increases. The improved performance of OTFS over OFDM remains unabashed for higher Dpt as well. Thus, the superiority of OTFS over OFDM by virtue of its diversity gain is clearly established.

It can also be found that OFDM is severely limited by residual synchronisation errors, which is not a new finding. In comparison we find that CP-OTFS has significant resilience to such large residual synchronization error although not completely immune to it.

Thus, it can be understood that the time domain channel estimation provides sufficient resilience to OTFS against synchronization errors while compensating for TVMC.

6.3 Conclusions

From the results presented in this chapter, one can conclude that block OFDM outperforms VSB-OFDM. It is also seen that in spite of using the same receiver as block OFDM, OTFS outperforms block OFDM. The only difference between block OFDM and OTFS is the time-interleaving of the transmitted signal which provides a gain of 2 dB. Overall, OTFS has 5 dB gain over VSB-OFDM for 16 QAM in LTV channel with FEC.

It is also see that the channel estimation in OTFS is less sensitive to estimation error, thus, leading to lower the pilot power in OTFS than OFDM.

Furthermore, the percentage resource used for pilots and CP overhead in OTFS is less than OFDM, and hence, the spectral efficiency (SE) of OTFS is higher than OFDM.

It was established that integer time and frequency errors result in a cyclic shift in delay and Doppler dimensions for CP-OTFS systems.

It was also shown that fractional Doppler or delay causes interference in Doppler or delay dimension, respectively.

It was found that the effect of synchronization errors can be considered as a part of the channel itself, however, with modified channel taps.

A time domain channel estimation method for De-Do domain embedded pilot based CP-OTFS system to estimate the effective channel matrix was described.

It is brought out that the time domain processing offers lower complexity owing to higher sparsity compared the De-Do domain channel in the presence of residual synchronization errors.

It was seen that CP-OTFS has significantly large tolerance to residual CFO, however, it is not completely immune to it.

The superiority of SIC algorithm was established, thus, recommending its use with OTFS systems. It was observed that SIC has a performance which is close to LMMSE with ideal channel estimates.

The massive gains especially for higher order QAM modulations that SIC provides are noteworthy.

The performance gain of CP-OTFS over OFDM by 3-7 dB is important to note.

6.3.1 Proof of Theorem 1

Proof. $y(k', L_p + l') = \sum_{l=0}^{\tilde{l}_\tau} \delta(l' - l) \sum_{k \in k_{\nu_i}} \left\{ \tilde{h}_{l,k} \Psi(K_p - k' + k_p, N) e^{j2\pi \frac{k_p(L + l' + L_p - l)}{(M+L)N}} \right\}$

. At lth tap, i.e., when $l = l'$, $y(k', L_p + l) = \sum_{k \in k_{\nu_i}} \left\{ \tilde{h}_{l,k} e^{j2\pi \frac{k(L_p + L)}{(M+L)N}} \Psi(K_p - k' + k, N) \right\}$.
When IFFT of (M+L)N point is applied and it can be shown that,

$$\sum_{k'=0}^{N-1} y(k', L_p + l) e^{j2\pi \frac{nk'}{N(M+L)}}$$

$$= \sum_{k \in k_{\nu_l}} \left\{ \tilde{h}_{l,k} e^{j2\pi \frac{k(L_p + L)}{(M+L)N}} \left(\frac{1}{N} \sum_{k_t=0}^{N-1} e^{j2\pi \frac{k_t(K_p + k)}{N}} \right. \right.$$

$$\left. \left. \sum_{k'=0}^{N-1} e^{j2\pi \frac{k'(n - k_t(M+L))}{N(M+L)}} \right) \right\}$$

$$
= \begin{cases} \sum_{k \in k_{\nu_l}} \left\{ \tilde{h}_{l,k} e^{j2\pi \frac{k(L_p+L)}{(M+L)N}} \left(e^{j2\pi \frac{n(K_p+k)}{N(M+L)}} \right) \right\}, n = \alpha(M+L) \\[2mm] \sum_{k_t=0}^{N-1} e^{j2\pi \frac{k_t(K_p)}{N}} \Psi((\frac{n}{(M+L)} - k_t), N) \sum_{k \in k_{\nu_l}} \\[2mm] \left\{ \tilde{h}_{l,k} e^{j2\pi \frac{k(L_p+L)}{(M+L)N}} \right\} e^{j2\pi \frac{k_t(k)}{N}}, \qquad otherwise. \end{cases}
$$

From (6.37), $h(n,m) =$

$$
\begin{cases} h(n + L_p + L + l, l) e^{j2\pi \frac{n(K_p)}{N(M+L)}}, \qquad n = \alpha(M+L) \\[2mm] \sum_{k_t=0}^{N-1} e^{j2\pi \frac{k_t(K_p)}{N}} \Psi((\frac{n}{(M+L)} - k_t), N) \\[2mm] h(k_t(M+L) + L_p + L + l, l), \qquad otherwise. \end{cases}
$$

Hence, $h(\alpha(M+L) + L + L_p + l, l) =$

$$
e^{-j2\pi \frac{\alpha K_p}{N}} \left(\sum_{k'=0}^{N-1} y(k', L_p + l) e^{j2\pi \frac{\alpha k'}{N}} \right)
$$

$\square$

6.3.2 Proof of Theorem 2

Proof. We know, $\hat{h}(\alpha(M+L) + L_p + l, l) = e^{-j2\pi \frac{\alpha K_p}{N}} \left(\sum_{k'=0}^{N-1} y(k', L_p + l) e^{j2\pi \frac{\alpha k'}{N}} \right)$.
Also, as explained in the [118], the time domain signal can be viewed as the interleaved OFDM, in which an N-point FFT is taken along the Doppler axis and then interleaved to get the time domain signal. If this is applied to the received De-Doh signal $y(k', l')$, then the signal without CP can be given as, $r(\alpha(M) + l') = \left(\sum_{k'=0}^{N-1} y(k', l') e^{j2\pi \frac{\alpha k'}{N}} \right)$.
Due to addition of CP, $r(\alpha(M+L) + L + l') = \left(\sum_{k'=0}^{N-1} y(k', l') e^{j2\pi \frac{\alpha k'}{N}} \right)$. Therefore,
$\hat{h}(\alpha(M+L) + L_p + l, l) = e^{-j2\pi \frac{\alpha K_p}{N}} r(\alpha(M+L) + L + L_p + l)$. $\square$

6.3.3 PROOF: Delay-Doppler Input-Output Relation

$$
y(k', l') = \frac{1}{\sqrt{NM}} \sum_{n'=0}^{N-1} \sum_{m'=0}^{M-1} Y(n', m') e^{-j2\pi [\frac{n'k'}{N} - \frac{m'l'}{M}]}
$$

Using (3.1) and (6.16), with some algebraic manipulations it can be simplified as,

$$
y(k', l') = \frac{1}{NM} \sum_{p=1}^{P} \tilde{h}_p e^{j2\pi \tilde{\nu}_p T_s (L - \frac{\tilde{\tau}_p}{T_s})}
$$

$$\sum_{k=0}^{N-1} \sum_{l=0}^{M-1} d(k,l) \sum_{m=0}^{M-1} \sum_{m'=0}^{M-1} e^{-j2\pi \frac{m\tilde{\tau}_p}{MT_s}} e^{j2\pi \frac{m'l'}{MT_s}} \tag{6.47}$$

$$\Psi(m + \tilde{\nu}_p T_s - m', M)$$

$$\sum_{n'=0}^{N-1} e^{j2\pi \frac{n'(\tilde{\nu}_p T_s N(M+L)+k-k')}{N}}. \text{ By substituting,}$$

$\Psi(m + \tilde{\nu}_p T_s - m', M) = \frac{1}{M} \sum_{k_t=0}^{M-1} e^{j2\pi \frac{k_t(m+\tilde{\nu}_p T_s - m')}{M}}$, we can write,

$$y(k',l') = \frac{1}{NM^2} \sum_{p=1}^{P} \tilde{h}_p e^{j2\pi \tilde{\nu}_p T_s (L - \frac{\tilde{\tau}_p}{T_s})}$$

$$\sum_{k=0}^{N-1} \sum_{l=0}^{M-1} d(k,l) \left(\sum_{n'=0}^{N-1} e^{j2\pi \frac{n'(\tilde{\nu}_p T_s N(M+L)+k-k')}{N}} \right) \tag{6.48}$$

$$\sum_{k_t=0}^{M-1} e^{j2\pi k_t(\tilde{\nu}_p T_s)} \sum_{m=0}^{M-1} e^{j2\pi \frac{m}{M}(k_t - l - \frac{\tilde{\tau}_p}{T_s})} \sum_{m'=0}^{M-1} e^{-j2\pi \frac{m'(l'-k_t)}{M}}$$

Since, $(l' - k_t) \in \mathbb{Z}$, we substitute $e^{-j2\pi \frac{m'(l'-k_t)}{M}} = M\delta[(l' - k_t)_M]$, $\frac{1}{N} \sum_{n'=0}^{N-1} e^{j2\pi \frac{n'(\tilde{\nu}_p T_s N(M+L)+k-k')}{N}} = \Psi(\tilde{\nu}_p T_s N(M+L) + k - k', N)$ and $\frac{1}{M} \sum_{m=0}^{M-1} e^{j2\pi \frac{m}{M}(k_t - l - \frac{\tilde{\tau}_p}{T_s})}$ $= \Psi(k_t - l - \frac{\tilde{\tau}_p}{T_s}, M)$, therefore, we get,

$$y(k',l') = \frac{1}{M} \sum_{p=1}^{P} \tilde{h}_p e^{j2\pi \tilde{\nu}_p T_s (L - \frac{\tilde{\tau}_p}{T_s})} \sum_{k=0}^{N-1} \sum_{l=0}^{M-1} d(k,l)$$

$$\Psi(\tilde{\nu}_p T_s N(M+L) + k - k', N) \sum_{k_t=0}^{M-1} e^{-j2\pi k_t(\tilde{\nu}_p T_s)}$$

$$\Psi(k_t - l - \frac{\tilde{\tau}_p}{T_s}, M) \left(M\delta[(l' - k_t)_M] \right). \tag{6.49}$$

Using $\Delta\tau = \frac{1}{B} = T_s$ $\Delta\nu = \frac{1}{N(M+L)T_s}$ and $l' = k_t$,

$$y(k',l') = \sum_{p=1}^{P} \tilde{h}_p e^{j2\pi \tilde{\nu}_p T_s(L)} \sum_{k=0}^{N-1} \sum_{l=0}^{M-1} d(k,l) \Psi(\frac{\tilde{\nu}_p}{\Delta\nu}$$

$$+ k - k', N) \Psi(l' - \frac{\tilde{\tau}_p}{\Delta\tau} - l, M) e^{-j2\pi(l' - \frac{\tilde{\tau}_p}{\Delta\tau})(\frac{\tilde{\nu}_p}{\Delta\nu N(M+L)})}$$

7

Nonorthogonal Multiple Access with OTFS

We discuss nonorthogonal multiple access for OTFS technology in this chapter.

The multiple access (MA) technique can be broadly classified in two categories, namely (i) orthogonal multiple access (OMA) and (ii) nonorthogonal multiple access (NOMA). In OMA, resource allocation orthogonality is maintained i.e., one resource unit is allocated to only one user. With reference to OTFS, two types of OMA-OTFS are known (i) TF MA OTFS [119]: where users are allocated different TF resources and (ii) De-Do MA-OTFS [120]: where users are allocated different De-Do resources.

In opposition to OMA, more than one user is allocated the same radio resource unit in NOMA. The method power-domain NOMA (PD-NOMA) is realized using superposition coding (SC) at the transmitter along with successive interference cancellation (SIC) at the receiver. Such schemes are known achieve the capacity of Gaussian broadcast channel. It is known that PD-NOMA outperform OMA as well as code-division NOMA schemes [121] in terms of sum spectral efficiency (SE) performance [122, 123].

7.1 OTFS Signal Model

The OTFS system model followed here is as in Chapter 3. However, as multiple users are invoved here, who transmit OTFS signal simultaneously therefore the expression now include user index i. The QAM modulated De-Do data symbols, $d_i(k, l) \in \mathbb{C}$, $k \in \mathbb{N}[0 \ N - 1]$, $l \in \mathbb{N}[0 \ M - 1]$, are arranged over De-Do lattice $\Lambda = \{(\frac{k}{NT}, \frac{l}{M\Delta f})\}$. Data symbols $d_i(k, l)$ is mapped to TF domain data $X_i(n, m)$ on lattice $\Lambda^\perp = \{(nT, \ m\Delta f)\}$, $n \in \mathbb{N}[0 \ N - 1]$ and $m \in \mathbb{N}[0 \ M - 1]$ by using inverse symplectic fast Fourier transform (ISFFT). Thus $X_i(n, m)$ can be given as [124],

$$X_i(n, m) = \frac{1}{\sqrt{NM}} \sum_{k=0}^{N-1} \sum_{m=0}^{M-1} d_i(k, l) e^{j2\pi[\frac{nk}{N} - \frac{ml}{M}]}. \tag{7.1}$$

169

Next, a TF modulator modulates $X_i(n, m)$ to time domain using Heisenberg transform as,

$$s_i(t) = \sum_{n=0}^{N-1} \sum_{m=0}^{M-1} (\sqrt{\mathscr{P}} X_i(n, m)) g(t - nT) e^{j2\pi m \Delta f(t-nT)}, \qquad (7.2)$$

where, $g(t)$ is transmitter pulse of duration T and transmit power is denoted by $\mathscr{P}$. Further, $s_i(t)$ is sampled at the sampling interval of $\frac{T}{M}$. We collect samples of $s_i(t)$ in $\mathbf{s}_i = [s_i(0)\ s_i(1) \cdots s_i(MN-1)]$. The QAM symbols $d_i(k, l)$ are arranged in $M \times N$ matrix as,

$$\mathbf{D}_i = \begin{bmatrix} d_i(0,0) & d_i(1,0) & \cdots & d_i(N-1,0) \\ d_i(0,1) & d_i(1,1) & \cdots & d_i(N-1,1) \\ \vdots & \vdots & \ddots & \vdots \\ d_i(M-1,0) & d_i(M-1,1) & \cdots & d_i(N-1,M-1) \end{bmatrix} \qquad (7.3)$$

The transmitted signal can be written as matrix-vector multiplication as:

$$\mathbf{s}_i = \mathbf{A}\sqrt{\mathscr{P}}\mathbf{d}_i, \qquad (7.4)$$

where, $\mathbf{d}_i = vec\{\mathbf{D}_i\}$. Finally, $\mathbf{A}_{MN \times MN} = \mathbf{W}_N \otimes \mathbf{I}_M$ denotes the OTFS modulation matrix. A cyclic prefix (CP) of length $L'_{CP} \geq L_{CP} - 1$ is appended at the starting of the $\mathbf{s}$, where L_{CP} is the channel's maximum excess delay length. In order to implement OFDM in the same framework, the modulation matrix is modified as $\mathbf{A} = \mathbf{I}_N \otimes \mathbf{W}_M$.

7.2 Delay-Doppler Power-Domain NOMA-OTFS

We consider K users multiplexed in power domain all of which are served by OTFS in both downlink and uplink transmission.

7.2.1 De-Do PD-NOMA-OTFS Downlink

7.2.1.1 Transmit Signal Model

Among the K users multiplexed in power domain, let the i-th user be allocated β_i fraction of total power $\mathscr{P}$, such that $\sum_{i=1}^{K} \beta_i = 1$. The methods to choose β_i are described in Sec. 7.3. Following the principle of superposition, the composite transmitted signal from the transmitter intended for all users can be written by modifying (7.4) as:

$$\mathbf{s} = \mathbf{A} \sum_{i=1}^{K} \sqrt{\beta_i \mathscr{P}} \mathbf{d}_i. \qquad (7.5)$$

Let, the i-th user's channel consists of P_i paths with h_{p^i} complex attenuations, τ_{p^i} delays and ν_{p^i} Doppler values for p^ith path where $p^i \in \mathbb{N}[1\ P_i]$. Thus, De-Do channel spreading function for the i-th user is given as,

$$h_i(\tau, \nu) = \sum_{p^i=1}^{P_i} h_{p^i} \delta(\tau - \tau_{p^i}) \delta(\nu - \nu_{p^i}), \ i = 1, \cdots, K. \tag{7.6}$$

The delay and Doppler values for p^ith path are $\tau_{p^i} = \frac{l_p^i}{M \Delta f}$ and $\nu_{p^i} = \frac{k_p^i}{NT}$, where $l_p^i \in \mathbb{N}[0 \ M-1]$ and $k_p^i \in \mathbb{N}[0 \ N-1]$ are the number of delay and Doppler bins on the De-Do lattice for p^ith path. N and M are assumed to be large enough so that one may neglect the effect of fractional delay and Doppler on the performance. In this part, perfect knowledge of $(h_{p^i}, l_{p^i}, k_{p^i})$, $p^i \in \mathbb{N}[0 \ P_i - 1]$, at the receiver of i-th user, is considered. Let τ_{max}^i and ν_{max}^i be the maximum delay and Doppler spread for users. Channel delay length $\alpha^i = \lceil \tau_{max}^i M \Delta f \rceil$ and channel Doppler length, $\beta^i = \lceil \nu_{max}^i NT \rceil$. $L_{CP} = \max\limits_{i=1,\cdots,K} (\alpha^i)$.

After removing CP, signal received at the i-th user's receiver, is [92],

$$\mathbf{r}_i = \mathbf{H}_i \mathbf{s} + \mathbf{n}_i, \ i = 1, \cdots, K. \tag{7.7}$$

where, $\mathbf{n}_i$ is white Gaussian noise vector of length MN with elemental variance $\sigma_{\mathbf{n}}^2$ and $\mathbf{H}_i$ is a $MN \times MN$ channel matrix for i^{th} user which is given by,

$$\mathbf{H}_i = \sum_{p^i=1}^{P_i} h_p^i \Pi^{l_p^i} \Delta^{k_p^i}, \ i = 1, \cdots, K, \tag{7.8}$$

with $\Pi_{MN \times MN} = circ\{[0 \ 1 \ 0 \cdots 0]_{MN \times 1}^T\}$ is a circulant delay matrix and $\Delta = diag[1 \ e^{j2\pi \frac{1}{MN}} \cdots e^{j2\pi \frac{MN-1}{MN}}]$ is a diagonal Doppler matrix.

7.2.1.2 Receiver Processing, SINR and SE Analysis

The receiver is considered to be made of LMMSE architecture discussed in previous chapters as the first stage, which is followed by code word level SIC in order to mitigate the NOMA interference. The total effective noise at the i-th receiver amounts to:

$$\tilde{\mathbf{n}}_{i_{DL}} = \sum_{i'=1, i' \neq i}^{K} \sqrt{\beta_{i'} \mathscr{P}} \mathbf{H}_i \mathbf{A} \mathbf{d}_{i'} + \mathbf{n}_i. \tag{7.9}$$

Assuming the total effective noise follows Gaussian distribution, LMMSE equalization on the received signal $\mathbf{r}_i$ in (7.7) results in estimated data vector for i-th user as given in (7.10), where, Γ_i denotes the average SNR of i-th user.

$$\hat{\mathbf{d}}_i = \sqrt{\beta_i} (\mathbf{H}_i \mathbf{A})^\dagger \left[\beta_i (\mathbf{H}_i \mathbf{A})(\mathbf{H}_i \mathbf{A})^\dagger + \sum_{i'=1, i' \neq i}^{K} \beta_{i'} (\mathbf{H}_i \mathbf{A})(\mathbf{H}_i \mathbf{A})^\dagger + \frac{1}{\Gamma_i} \mathbf{I} \right]^{-1} \mathbf{r}_i$$

$$= \sqrt{\beta_i}(\mathbf{H}_i\mathbf{A})^\dagger[(\mathbf{H}_i\mathbf{A})(\mathbf{H}_i\mathbf{A})^\dagger + \frac{1}{\Gamma_i}\mathbf{I}]^{-1}\mathbf{r}_i, \tag{7.10}$$

Rewriting (7.10) by using (7.5) and (7.7), one obtains:

$$\hat{\mathbf{d}}_i = \underbrace{\mathbf{B}_i\sqrt{(\beta_i\mathscr{P})}\mathbf{d}_i}_{\text{desired signal}} + \underbrace{\sum_{i'=1,i'\neq i}^{K} \mathbf{B}_i\sqrt{(\beta_{i'}\mathscr{P})}\mathbf{d}_{i'}}_{\text{NOMA interference}} \tag{7.11}$$

$$+ \underbrace{\mathbf{C}_i\mathbf{n}_i}_{\text{noise component}} , i = 1, \cdots, K,$$

where, for notational simplicity, $\mathbf{C}_i = \sqrt{\beta_i}(\mathbf{H}_i\mathbf{A})^\dagger[(\mathbf{H}_i\mathbf{A})(\mathbf{H}_i\mathbf{A})^\dagger + \frac{1}{\Gamma_i}\mathbf{I}]^{-1}$ and $\mathbf{B}_i = \mathbf{C}_i\mathbf{H}_i\mathbf{A}$ are assigned. Without loss of generality, one can consider that the i-th user is at a higher distance from the transmitting BS than the $(i+1)$-th user for $i = 1, \cdots, (K-1)$. This can be translated to received average SNR, as: $\Gamma_1 < \Gamma_2 < \cdots < \Gamma_{i-1} < \Gamma_i < \cdots < \Gamma_K$. Thus, we assume that following the principle of NOMA, the i-th user will not face any interference due to the signals intended for 1st, 2nd,$\cdots$, $(i-1)$-th users through the use of perfect SIC receiver processing. Using these assumptions and expanding (7.11), the symbol-wise pre and post SIC received SINR at any user can be formulated. For the i-th user, the downlink pre and post SIC SINR for j-th symbol (denoted as $\Upsilon_{ij}^{\text{Pre-D}}$ and $\Upsilon_{ij}^{\text{Post-D}}$, respectively) can be given by (7.12) and (7.13) respectively, with $i = 1, \cdots, K$ and $j = 1, \cdots, MN$. b_{pq}^i and c_{pq}^i denote the $(p,q)^{\text{th}}$ elements of $\mathbf{B}_i$ and $\mathbf{C}_i$, respectively.

$$\Upsilon_{ij}^{\text{Pre-D}} = \frac{\overbrace{\beta_i\mathscr{P}|b_{jj}^i|^2}^{\text{desired power}}}{\left[\underbrace{\beta_i\mathscr{P}\sum_{l=1,l\neq j}^{MN}|b_{jl}^i|^2}_{\text{inter-symbol interference}} + \underbrace{\sum_{i'=1,i'\neq i}^{MN}\beta_{i'}\mathscr{P}(\sum_{l=1}^{MN}|b_{jl}^i|^2)}_{\text{NOMA interference}} + \underbrace{\sum_{l=1}^{MN}|c_{jl}^i|^2\sigma_n^2}_{\text{noise power}}\right]}, \tag{7.12}$$

$$\Upsilon_{ij}^{\text{Post-D}} = \frac{\overbrace{\beta_i\mathscr{P}|b_{jj}^i|^2}^{\text{desired power}}}{\left[\underbrace{\beta_i\mathscr{P}\sum_{l=1,l\neq j}^{MN}|b_{jl}^i|^2}_{\text{inter-symbol interference}} + \underbrace{\sum_{i'=i+1}^{MN}\beta_{i'}\mathscr{P}(\sum_{l=1}^{MN}|b_{jl}^i|^2)}_{\text{NOMA interference}} + \underbrace{\sum_{l=1}^{MN}|c_{jl}^i|^2\sigma_n^2}_{\text{noise power}}\right]}, \tag{7.13}$$

Since, the SINR achieved in all symbols of OTFS are nearly same for large values of M and N, [125], the subscript i is dropped. The pre and post SIC SINR of i-th user are $\Upsilon_i^{\text{Pre-D}}$ and $\Upsilon_i^{\text{Post-D}}$, respectively. Thus, the downlink sum rate of the system in bps/Hz is given by:

$$R_{sum}^{DL} = \sum_{i=1}^{K}\log_2(1 + \Upsilon_i^{\text{Post-D}}). \tag{7.14}$$

It is noteworthy that the SE performance presented here for downlink (and subsequently for uplink in Sec 7.2.2.2) are done for such realizable MMSE-SIC receiver only. SE calculation using log-determinant method[1] of the De-Do channel $\mathbf{H}_i$ is not considered here.

7.2.2 De-Do PD-NOMA-OTFS Uplink

7.2.2.1 Transmit Signal Model

In the case of uplink OTFS-NOMA, it is considered that all the K users transmit data simultaneously to the base station in De-Do domain, thus making it a multiple-access channel (MAC). Here, perfect carrier and clock synchronization among the transmitting users has been assumed. It is also considered assumed that all users have the same OTFS grid size (M, N). Perfect knowledge about user channels is asumed at the receiver of the base station. The OTFS modulated transmitted vector from i-th user is given by:

$$\mathbf{s}_i^u = \mathbf{A}\sqrt{\mathscr{P}_i^u}\mathbf{d}_i^u, \qquad (7.15)$$

where, $\mathscr{P}_i^u$ and $\mathbf{d}_i^u$ denote the transmit power and vectorized transmit data of the i-th user, respectively. The uplink average SNR of the i-th user is given by $\Gamma_i^u = \mathscr{P}_i^u/\sigma_n^2$. The aggregate received signal at the base station after removal of CP is given by:

$$\mathbf{r}_u = \sum_{i=1}^{K} \mathbf{H}_i^u \mathbf{s}_i^u + \mathbf{n}, \qquad (7.16)$$

where, $\mathbf{H}_i^u$ denotes the $MN \times MN$ De-Do uplink channel matrix from i-th user to the BS.

7.2.2.2 Receiver Processing, SINR and SE Analysis

During uplink transmission in NOMA, the signal from the users with higher SNR are sequentially decoded and successively canceled from the aggregate signal. For the same user ordering as considered in downlink transmission ($\Gamma_1^u < \cdots < \Gamma_K^u$), while decoding the i-th user's signal, the BS will consider the 1st, 2nd, $\cdots$, $(i-1)$-th users' signals as noise. Thus, for i-th user, the effective noise can be denoted as:

$$\tilde{\mathbf{n}}_i = \sum_{i'=1}^{i-1} \sqrt{\mathscr{P}_{i'}^u}\mathbf{H}_{i'}^u \mathbf{A}\mathbf{d}_{i'}^u + \mathbf{n}. \qquad (7.17)$$

Thus, the noise variance for the i-th user is given by:

$$\tilde{\sigma}_{n_i}^2 = \mathbb{E}[\tilde{\mathbf{n}}_i \tilde{\mathbf{n}}_i^H] = \sum_{i'=1}^{i-1} \mathscr{P}_{i'}^u \mathbf{H}_i \mathbf{H}_i^H + \sigma_n^2 \mathbf{I}. \qquad (7.18)$$

After processing the received signal ($\mathbf{r}_u$) through LMMSE equalizer (similar to (7.10) for downlink), the estimated data vector for the i-th user at the BS can be expressed by (7.19).

$$\hat{\mathbf{d}}_i^u = (\mathbf{H}_i^u \mathbf{A})^\dagger [(\mathbf{H}_i^u \mathbf{A})(\mathbf{H}_i^u \mathbf{A})^\dagger + \sum_{i'=1}^{i-1} \frac{\mathscr{P}_{i'}^u}{\mathscr{P}_i^u}(\mathbf{H}_i^u \mathbf{A})(\mathbf{H}_i^u \mathbf{A})^\dagger + \frac{\sigma_n^2}{\mathscr{P}_i^u}\mathbf{I}]^{-1}\mathbf{r}_i$$

---

[1] as usually done in conventional point-to-point multiple-input multiple-output (MIMO) systems [126]

$$= (\mathbf{H}_i^u \mathbf{A})^\dagger [\mathbf{H}_i^u \mathbf{H}_i^{u\dagger} + \sum_{i'=1}^{i-1} \frac{\Gamma_{i'}^u}{\Gamma_i^u} \mathbf{H}_{i'}^u \mathbf{H}_{i'}^{u\dagger} + \frac{1}{\Gamma_i^u} \mathbf{I}]^{-1} \mathbf{r}_i, \tag{7.19}$$

For notational simplicity, we denote $\mathbf{C}_i^u = (\mathbf{H}_i^u \mathbf{A})^\dagger [(\mathbf{H}_i^u \mathbf{A})(\mathbf{H}_i^u \mathbf{A})^\dagger + \sum_{i'=1}^{i-1} \frac{\Gamma_{i'}^u}{\Gamma_i^u} (\mathbf{H}_i^u \mathbf{A})$ $(\mathbf{H}_i^u \mathbf{A})^\dagger + \frac{1}{\Gamma_i^u} \mathbf{I}]^{-1}$, $\mathbf{B}_{ii}^u = \mathbf{C}_i^u \mathbf{H}_i^u \mathbf{A}$ and $\mathbf{B}_{ii'}^u = \mathbf{C}_i^u \mathbf{H}_{i'}^u \mathbf{A}$. Thus combining (7.15) and (7.16), (7.19) the following can be written:

$$\hat{\mathbf{d}}_i^u = \underbrace{\mathbf{B}_{ii}^u \sqrt{\mathscr{P}_i^u} \mathbf{d}_i^u}_{\text{desired signal}} + \underbrace{\sum_{i'=1}^{i-1} \mathbf{B}_{ii'}^u \sqrt{\mathscr{P}_{i'}^u} \mathbf{d}_{i'}}_{\text{NOMA interference}} + \underbrace{\mathbf{C}_i^u \mathbf{n}_i}_{\text{noise component}} , \quad i = 1, \cdots, K, \tag{7.20}$$

Expanding (7.20), the uplink SINR for j-th symbol of the i-th user can be written as:

$$\Upsilon_{ij}^{\mathrm{U}} = \frac{\overbrace{\mathscr{P}_i^u |b_{jj}^{uii}|^2}^{\text{desired power}}}{\left[\underbrace{\mathscr{P}_i^u \sum_{l=1, l \neq j}^{MN} |b_{jl}^{uii}|^2}_{\text{inter-symbol interference}} + \underbrace{\sum_{i'=1}^{i-1} \mathscr{P}_{i'}^u (\sum_{l=1}^{MN} |b_{jl}^{uii'}|^2) +}_{\text{NOMA interference}} \underbrace{\sum_{l=1}^{MN} |c_{jl}^{ui}|^2 \sigma_n^2}_{\text{noise power}} \right]}, \tag{7.21}$$

with $i = 1, \cdots, K$ and $j = 1, \cdots, MN$. b_{pq}^{uii}, $b_{pq}^{uii'}$ and c_{pq}^{ui} denote the $(p, q)^{\text{th}}$ elements of $\mathbf{B}_{ii}^u$, $\mathbf{B}_{ii'}^u$ and $\mathbf{C}_i^u$, respectively. Using similar assumptions made for downlink direction, the sum rate (in bps/Hz) in uplink direction is given by:

$$R_{sum}^{UL} = \sum_{i=1}^{K} \log_2(1 + \Upsilon_i^{\mathrm{U}}). \tag{7.22}$$

7.3 Power Allocation Schemes Among Download NOMA-OTFS Users

7.3.1 Fixed Power Allocation (FPA)

FPA is a static power allocation method. Here, the fractions of total transmit power allocated for different users are predetermined. Such conventional scheme has been used in NOMA performance analysis for simplicity and in order to have a benchmark for other sophisticated power allocation strategies. The power fractions are independent of user channel conditions and system performance. In order to maintain fairness among users, it is a general practice to allocate more power to the users with lower received average SNR. Thus, for the SNR order mentioned in 7.2.1.2, the fixed transmit power fractions will be ordered as: $\beta_1 > \beta_2 \cdots > \beta_K$, with the constraint $\sum_{i=1}^{K} \beta_i = 1$.

7.3.2 Fractional Transmit Power Allocation (FTPA)

This is a dynamic power allocation scheme. The fraction of power allocated to a user is made proportional to the inverse of its channel gain. Thus, the users with lower channel gain is allocated higher transmit power in order to maintain system fairness. The following two FTPA schemes are described which have different level of channel state information from the user links.

7.3.2.1 Average SNR Based FTPA
Here, it is assumed that the BS has the access to only slowly time-varying average received SNR values (Γ_i) of the users. The fraction of power allocated to the i-th user is given by:

$$\beta_i = \frac{\Gamma_i^{-1}}{\sum_{i'=1}^{K} \Gamma_{i'}^{-1}}. \tag{7.23}$$

7.3.2.2 Channel Norm Based FTPA
In this scheme the scenario, where the base station has access to the partial CSI of all users in terms of the instantaneous channel norms is considered. The fraction of power allocated to the i-th user is given by:

$$\beta_i = \frac{||\mathbf{H}_i||_F^{-1}}{\sum_{i'=1}^{K} ||\mathbf{H}_{i'}||_F^{-1}}, \tag{7.24}$$

where, $\mathbf{H}_i$ is defined in (7.8).

7.3.3 Power Allocation for Weighed Sum Rate Maximization (WSRM)

Similar to FTPA, two weighted sum rate maximization frameworks based on average SNR information and instantaneous channel information at the base station are described.

7.3.3.1 Average SNR Based WSRM
When the base station has access to the average SNR information of the users, the optimization problem can be formulated based on the AWGN rates as described below:

$$\text{Maximize} \quad R_{sum}^{AWGN} = w_1 \log(1 + \frac{\beta_1 \Gamma_1}{1 + \beta_2 \Gamma_2}) + w_2 \log(1 + \beta_2 \Gamma_2),$$
$$\text{subject to} \quad \beta_1 + \beta_2 = 1, 0 \leq \beta_1, \beta_2 \leq 1; \tag{7.25}$$

where, w_1 and w_2 are the weights assigned to the two users in order to maintain fairness in power allocation. A suboptimal solution of the maximization problem is obtained by differentiating the cost function, as done in [127, Sec. III.A], although it is not straightforward to show concavity of such cost function in (7.25) . Reducing the problem in terms of only β_2, differentiating R_{sum}^{AWGN} w.r.t. β_2 and equating it to zero finally yields:

$$\frac{w_1 \Gamma_2}{1 + \beta_2 \Gamma_2} - \frac{w_2 \Gamma_1}{1 + \beta_2 \Gamma_1} = 0. \tag{7.26}$$

Solving the linear equation, the optimal value of β_2 can be obtained as:

$$\beta_2^{Opt} = \frac{w_2\Gamma_1 - w_1\Gamma_2}{(w_1 - w_2)\Gamma_1\Gamma_2}. \tag{7.27}$$

In order to impose the associated constraints stated in (7.25), we assign $\beta_2^{Opt} = \max(0, \min(1, \beta_2^{Opt}))$. Clearly, $\beta_1^{Opt} = 1 - \beta_2^{Opt}$.

7.3.3.2 Instantaneous Channel Information Based WSRM

If the base station has access to partial information about the instantaneous channel of each user (in terms of $\mathbf{B}_i$ and $\mathbf{C}_i$ matrices defined after (7.10)), the 2-user optimization problem can be formulated using the exact post SIC SINR expression (for j-th symbol) derived in (7.13) as follows:

$$\text{Maximize } R_{sum}^{Inst} = w_1\log(1 + \Upsilon_{1j}^{\text{Post-D}}) + w_2\log(1 + \Upsilon_{2j}^{\text{Post-D}})$$
$$\text{subject to } \beta_1 + \beta_2 = 1, 0 \le \beta_1, \beta_2 \le 1. \tag{7.28}$$

Using the notations $\mathscr{P}|b_{jj}^1|^2 = \Gamma_{1d}$, $\mathscr{P}\sum_{l=1,l\neq j}^{MN}|b_{jl}^1|^2 = \Gamma_{1\text{ISI}}$, $\mathscr{P}(\sum_{l=1}^{MN}|b_{jl}^1|^2) = \Gamma_{1\text{N}}$, $\sigma_n^2\sum_{l=1}^{MN}|c_{jl}^1|^2 = P_{1n}$, $\mathscr{P}|b_{jj}^2|^2 = \Gamma_{1d}$, $\mathscr{P}\sum_{l=1,l\neq j}^{MN}|b_{jl}^2|^2 = \Gamma_{2\text{ISI}}$, and $\sigma_n^2\sum_{l=1}^{MN}|c_{jl}^2|^2 = P_{2n}$, the instantaneous weighted sum rate in terms of β_2 can be expressed as:

$$R_{sum}^{Inst} = w_1\log[\frac{(1-\beta_2)(\Gamma_{1d} + \Gamma_{1\text{ISI}}) + \beta_2\Gamma_{1\text{N}} + P_{1n}}{(1-\beta_2)\Gamma_{1\text{ISI}} + \beta_2\Gamma_{1\text{N}} + P_{1n}}] + \tag{7.29}$$
$$w_2\log[\frac{\beta_2(\Gamma_{2d} + \Gamma_{2\text{ISI}}) + P_{2n}}{\beta_2\Gamma_{2\text{ISI}} + P_{2n}}].$$

$$\frac{w_1(\Gamma_{1\text{N}} - \Gamma_{1\text{ISI}} - \Gamma_{1d})}{(1-\beta_2)(\Gamma_{1d} + \Gamma_{1\text{ISI}}) + \beta_2\Gamma_{1\text{N}} + P_{1n}} - \frac{w_1(\Gamma_{1\text{N}} - \Gamma_{1\text{ISI}})}{(1-\beta_2)\Gamma_{1\text{ISI}} + \beta_2\Gamma_{1\text{N}} + P_{1n}}$$
$$+ \frac{w_2(\Gamma_{2d} + \Gamma_{2\text{ISI}})}{\beta_2(\Gamma_{2d} + \Gamma_{2\text{ISI}}) + P_{2n}} - \frac{w_2\Gamma_{2\text{ISI}}}{\beta_2\Gamma_{2\text{ISI}} + P_{2n}} = 0 \tag{7.30}$$

As done in previous section, we use the differentiation method in order to obtain a suboptimal solution of β_2 [127, Sec. III.A]. Differentiating R_{sum}^{Inst} w.r.t. β_2 and equating it to zero results in (7.30). By numerically solving (7.30) using available software tools, optimal value of β_2 (β_2^{Opt}) can be obtained. Similar to the earlier case, we finally assign $\beta_2^{Opt} = \max(0, \min(1, \beta_2^{Opt}))$ and $\beta_1^{Opt} = 1 - \beta_2^{Opt}$.

It is to be noted that judicious assignment of weights for the users has been addressed in literature considering proportional fairness [128, 127]. However, for simplicity we consider assignment of fixed weights as: $w_1 = 0.6, w_2 = 0.4$.

It is also important to note that in order to implement the power allocation schemes described in Sec. 7.3.2.2 and 7.3.3.2, channel information (like channel norm for the first scheme and $\mathbf{B}$ and $\mathbf{C}$ matrices for the second scheme) have to be measured and feedback at least once in every De-Do coherence time of the channel. However, it is known the De-Do coherence time of the channel is significantly larger than the coherence time in TF domain for OFDM [129]. Thus, instantaneous De-Do CSI-based NOMA power allocation schemes in high Doppler scenarios are realizable.

7.4 Link Level Performance Analysis of NOMA-OTFS Systems

Here, we consider 2 user multiplexed system as in [130, 131, 132].

7.4.1 Downlink MMSE SIC Receiver with LDPC Coding

Here, we describe a practical realization of a 2-user LDPC enabled codeword level SIC OTFS-NOMA receiver for downlink transmission as in Figure 7.1. The base station generates the data for both the users (denoted as b_1 and b_2, respectively), encode using the LDPC encoder and then modulate the data using modulation supported by the user. The encoded signals for both the users are denoted as d_1 and d_2, respectively.The modulated symbols are superimposed with allocated power (β_i). The superimposed time-domain signal is then OTFS modulated using SFFT and the Heisenberg transform. The resulting signal s is broadcast through the De-Do channel to both the users.

Here, we we let $K = 2$ in (7.7). Both the users first perform LMMSE equalization in order to mitigate the ISI and ICI. It is also assumed that SNR of second user is higher than the first user, thus, leading the second user to perform SIC operation.

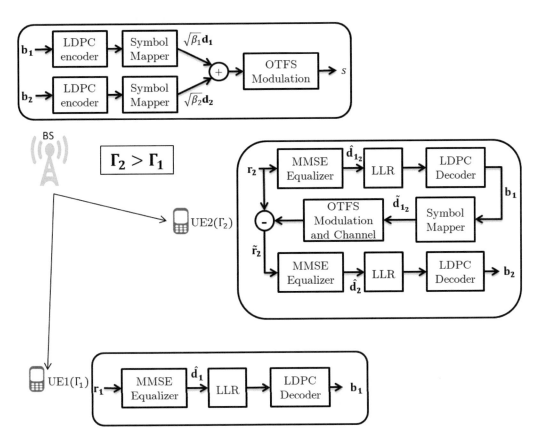

Figure 7.1 Representative block diagram of 2-user NOMA-OTFS system in downlink.

7.4.1.1 Processing at First User

The MMSE equalized data can be described using (7.10) with $i = 1$. In order to LDPC decode the equalized data, the channel LLR values are calculated from the equalized symbols as,

$$\mathscr{L}(b^j_{1\eta}|\hat{\mathbf{d}}_1(\eta)) \approx (\min_{s \in S^0_j} \frac{||\hat{\mathbf{d}}_1(\eta) - s||^2}{\sigma^2_1(\eta)}) - (\min_{s \in S^1_j} \frac{||\hat{\mathbf{d}}_1(\eta) - s||^2}{\sigma^2_1(\eta)}) \qquad (7.31)$$

where, $\mathbf{d}_i(\eta)$ is the ηth element of $\mathbf{d}_i$ mapped from the bits $b^0_{i\eta} \; b^1_{i\eta} \cdots b^{K_i-1}_{i\eta}$, K_i is the number of bits per symbol for user i and $\sigma^2_1(\eta)$ is the ηth element of $\sigma^2_1 = \frac{1}{\beta_1} diag(\sigma^2_n \mathbf{C}_1 \mathbf{C}^\dagger_1 + \beta_2 \mathbf{B}_1 \mathbf{B}^\dagger_1)$. The aggregate interference and noise is assumed to follow Gaussian distribution. S^k_j denotes the set of constellation symbols in which the bit $b^j = k$. These LLRs are then fed into the LDPC decoder to decode first user's data. Let $\mathbf{L}^1$ denotes a matrix where $\mathbf{L}^1(\eta, j) = \mathscr{L}(b^j_{1\eta}|\hat{\mathbf{d}}_1(\eta))$ for $\eta = 1, 2, \cdots, MN$ and $j = 0, 1, \cdots, K_i - 1$. $\mathbf{L}^1$ is reshaped to $L_{cl} \times N_{cw}$ matrix, where L_{cl} and N_{cw} denote the LDPC codeword, length and number of codewords respectively. Each column of $\mathbf{L}^1$, subsequently regenerates codeword c^1_ι for $\iota = 1, 2 \cdots, N_{cw}$ using the Min–Sum Algorithm [109] employed by the LDPC decoder. This algorithm iteratively updates the variable node and check node equation as discussed in earlier chapters.

7.4.1.2 Processing at Second User

Since it is assumed that second user experiences higher SNR, it performs SIC operation in which it decodes first user's data and then uses it to cancel the interference to decode its own data. The detected first user's data at the second user is given as,

$$\hat{\mathbf{d}}_{1_2} = \sqrt{\beta_1}(\mathbf{H}_2 \mathbf{A})^\dagger [(\mathbf{H}_2 \mathbf{A})(\mathbf{H}_2 \mathbf{A})^\dagger + \frac{1}{\Gamma_2}\mathbf{I}]^{-1} \mathbf{r}_2 \qquad (7.32)$$

Corresponding LLR of the equalized data of fist user is calculated as,

$$\mathscr{L}(b^j_{1_2\eta}|\hat{\mathbf{d}}_{1_2}(\eta)) \approx (\min_{s \in S^0_j} \frac{||\hat{\mathbf{d}}_{1_2}(\eta) - s||^2}{\sigma^2_{1_2}(\eta)}) - (\min_{s \in S^1_j} \frac{||\hat{\mathbf{d}}_{1_2}(\eta) - s||^2}{\sigma^2_{1_2}(\eta)}), \qquad (7.33)$$

where, $\sigma^2_{1_2}(\eta)$ is the ηth element of $\sigma^2_{1_2} = \frac{1}{\beta_1} diag(\sigma^2_n \mathbf{C}_2 \mathbf{C}^\dagger_2 + \beta_2 \mathbf{B}_2 \mathbf{B}^\dagger_2)$. The residual received signal at second user after canceling the interference due to first user is given by,

$$\tilde{\mathbf{r}}_2 = \mathbf{r}_2 - \sqrt{\beta_1 \mathscr{P}} \mathbf{H}_2 \mathbf{A} \tilde{\mathbf{d}}_{1_2}, \qquad (7.34)$$

where, $\tilde{\mathbf{d}}_{1_2}$ is generated at second user after passing the LDPC decoded codeword obtained from $\hat{\mathbf{d}}_{1_2}$ through symbol mapper. After MMSE equalization on the residual signal given in (7.34), the detected second user's data at the second user itself is given by,

$$\hat{\mathbf{d}}_2 = \sqrt{\beta_2}(\mathbf{H}_2 \mathbf{A})^\dagger [\beta_2(\mathbf{H}_2 \mathbf{A})(\mathbf{H}_2 \mathbf{A})^\dagger + \frac{1}{\Gamma_2}\mathbf{I}]^{-1} \tilde{\mathbf{r}}_2 \qquad (7.35)$$

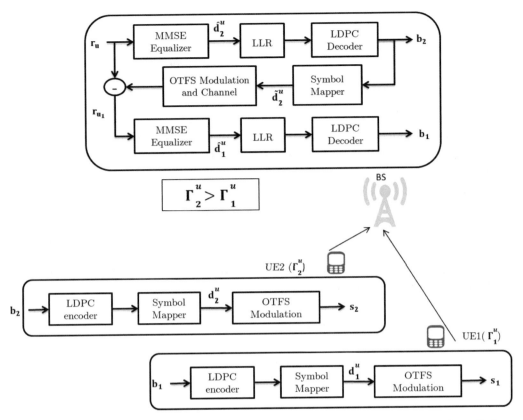

Figure 7.2 Representative block diagram of 2-user NOMA-OTFS system in uplink.

As for first user, the bit level LLRs for second user from the symbols are calculated as,

$$\mathscr{L}(b_{2\eta}^{j}|\hat{\mathbf{d}}_2(\eta)) \approx (\min_{s \in S_j^0} \frac{||\hat{\mathbf{d}}_2(\eta) - s||^2}{\sigma_2^2(\eta)}) - (\min_{s \in S_j^1} \frac{||\hat{\mathbf{d}}_2(\eta) - s||^2}{\sigma_2^2(\eta)}) \qquad (7.36)$$

where, $\sigma_2^2(\eta)$ is the ηth element of $\sigma_2^2 = \frac{1}{\beta_2} diag(\sigma_n^2 \mathbf{C}_2 \mathbf{C}_2^\dagger)$. The LLRs are updated and the data of user 2 is generated by the LDPC decoder to generate the data using Min–Sum algorithm as described earlier

7.4.2 Uplink MMSE SIC Receiver with LDPC Coding

The block diagrmatic representation of an uplink LDPC coded OTFS-NOMA for two users is depicted in Figure 7.2. For two user case, the received signal can be expressed using (7.16) using $K = 2$. Since it is assumed that $\Gamma_1^u < \Gamma_2^u$, we first decode second user's data at the base station as,

$$\hat{\mathbf{d}}_2^u = (\mathbf{H}_2^u \mathbf{A})^\dagger [(\mathbf{H}_2^u \mathbf{A})(\mathbf{H}_2^u \mathbf{A})^\dagger + \frac{\Gamma_1^u}{\Gamma_2^u}(\mathbf{H}_1^u \mathbf{A})(\mathbf{H}_1^u \mathbf{A})^\dagger + \frac{1}{\Gamma_2^u}\mathbf{I}]^{-1} \mathbf{r}_u. \qquad (7.37)$$

LLR values of second user can be computed as,

$$\mathscr{L}(b_{2\eta}^{u^j}|\hat{\mathbf{d}}_2^u(\eta)) \approx (\min_{s \in S_j^0} \frac{||\hat{\mathbf{d}}_2^u(\eta) - s||^2}{\boldsymbol{\sigma}_2^2(\eta)}) - (\min_{s \in S_j^1} \frac{||\hat{\mathbf{d}}_2^u(\eta) - s||^2}{\boldsymbol{\sigma}_2^2(\eta)}), \qquad (7.38)$$

where, η^{th} element of $\mathbf{d}_i^u$, $\mathbf{d}_i^u(\eta)$ is mapped from bits $b_{i\eta}^{u^0}\ b_{i\eta}^{u^1}\cdots b_{i\eta}^{u^{(K_i-1)}}$, K_i is the number of bits per symbol for user i and $\boldsymbol{\sigma}_2^2(\eta)$ is the ηth element of $\boldsymbol{\sigma}_2^2 = \frac{1}{\mathscr{P}_2} diag(\sigma_n^2 \mathbf{C}_2^u \mathbf{C}_2^{u\dagger} + \mathscr{P}_1 \mathbf{B}_{21}^u \mathbf{B}_{21}^{u\dagger})$. The calculated LLR values are processed by LDPC decoder in order to produce the message word. The obtained message is again encoded and modulated to generate the recovered data $\tilde{\mathbf{d}}_2^u$ for second user, which is used to cancel the interference from aggregate received signal to decode the first user's data as,

$$\tilde{\mathbf{r}}_{u_1} = \mathbf{r}_u - \sqrt{\mathscr{P}_2^u} \mathbf{H}_2^u \mathbf{A} \tilde{\mathbf{d}}_2^u. \qquad (7.39)$$

The signal at the BS after equalization of residual signal, using LMMSE method, is given by (7.39). The first user's data is given by,

$$\hat{\mathbf{d}}_1^u = (\mathbf{H}_1\mathbf{A})^\dagger [(\mathbf{H}_1\mathbf{A})(\mathbf{H}_1\mathbf{A})^\dagger + \frac{1}{\Gamma_1}\mathbf{I}]^{-1}\tilde{\mathbf{r}}_{u_1}. \qquad (7.40)$$

The equalized data $\hat{\mathbf{d}}_1^u$ is used to calculate the LLR as follows:

$$\mathscr{L}(b_{1\eta}^j|\hat{\mathbf{d}}_1^u(\eta)) \approx (\min_{s \in S_j^0} \frac{||\hat{\mathbf{d}}_1^u(\eta) - s||^2}{\boldsymbol{\sigma}_1^2(\eta)}) - (\min_{s \in S_j^1} \frac{||\hat{\mathbf{d}}_1^u(\eta) - s||^2}{\boldsymbol{\sigma}_1^2(\eta)}), \qquad (7.41)$$

where, assuming perfect SIC, $\hat{\mathbf{d}}_1^u(\eta)$ and $\boldsymbol{\sigma}_2^2(\eta)$ are the ηth element of $\hat{\mathbf{d}}$ and $\boldsymbol{\sigma}_1^2 = \frac{1}{\mathscr{P}_1^u} diag(\sigma_n^2 \mathbf{C}_1^u \mathbf{C}_1^{u\dagger})$. The calculated LLR values are then fed to LDPC decoder to reproduce the data of user 1.

7.5 Simulation Results and Discussion

The performance analysis of OTFS-NOMA in terms of both system level and link level evaluation for both downlink and uplink are discussed below. The important simulation parameters are listed in Table 7.1. An equivalent OFDM system with synchronous CP length and block based signal structure as described in Sec. 7.1 is considered for comparison.

The MMSE equalizer, shown in earlier chapters, can cancel the ICI at the receiver, unlike the single-tap equalizer used in traditional systems [134].

While 5G-NR has provision for variable subcarrier bandwidth [135, 136], the EVA channel model restricts the use of subcarrier bandwidth of upto 60 KHz corresponding to numerology 2 only since the the coherence bandwidth is about 56 KHz. Channel equalization with MMSE uses ideal channel estimates as described in the sections above. The system performance is then accordingly evaluated using a subcarrier bandwidth of 15 KHz which is also valid for system design of 4G systems [137, 138].

Table 7.1 Key system parameters.

Parameter	Value
LTV De-Do channel model	"Extended Vehicular A (EVA)"[133]
Doppler slots (N)	16
Delay slots (M)	256
Number of NOMA users	2
User speed	500 Kmph
Carrier frequency	5.9 GHz
Subcarrier bandwidth Δf	15 KHz
Total bandwidth B	3.84 MHz
Frame duration T_f	1.08 ms
Error correction codes	LDPC codes. code length = 648, code rate (R) = 2/3 [108]
Downlink average SNR	$\Gamma_1 = 15$ dB, $\Gamma_2 = 25$ dB
Uplink average SNR	$\Gamma_1^u = 10$ dB, $\Gamma_2^u = 30$ dB

The performance is evaluated in terms of metrics like BLER (P_e), throughput and goodput. The throughput of a link is defined as the number of bits transmitted per unit time and is given by

$$\text{Throughput} = \Sigma_{i=1}^2 \text{RK}_i \text{ bits/s/Hz}, \tag{7.42}$$

where, R and K_i denote code rate and bits per QAM symbol, respectively. The goodput of a link is defined as the number of bits that are successfully received and expressed as

$$\text{Goodput} = \Sigma_{i=1}^2 \text{RK}_i(1 - P_{e_i}) \text{ bits/s/Hz}, \tag{7.43}$$

where, P_{e_i} denotes BLER for i-th user.

7.5.1 System Level Spectral Efficiency Results

7.5.1.1 Comparison between NOMA/OMA-OTFS

In Figure 7.3a and 7.3b, the cumulative distribution functions (CDF) of downlink sum rates achieved under OMA and various NOMA power allocation schemes in case of OTFS transmission are shown for downlink average SNR values $(\Gamma_1, \Gamma_2) = (15, 25)$ dB and $(10, 30)$ dB, respectively. The exact values of mean and 5% outage sum rate are given in Table 7.2 for $(\Gamma_1, \Gamma_2) = (15, 25)$ dB. Mean and 5% outage sum rate for $(\Gamma_1, \Gamma_2) = (10, 30)$ dB are visually shown in Figure 7.4 and 7.5, respectively.

OMA-OTFS is taken as the benchmark. some power allocation schemes in the NOMA-OTFS scheme (namely fixed with chosen power ratio, channel norm based FTPA and instantaneous channel information- based WSRM) provide significant increase in mean and outage sum rate. For example, we notice that there is more than 16% improvement in both mean and 5% outage sum SE, respectively in case of NOMA-OTFS with average

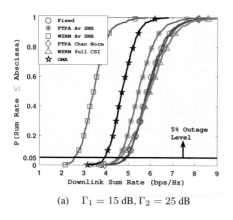 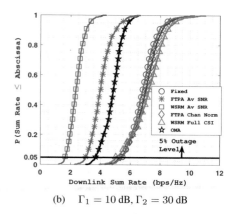

(a) $\Gamma_1 = 15$ dB, $\Gamma_2 = 25$ dB   (b) $\Gamma_1 = 10$ dB, $\Gamma_2 = 30$ dB

Figure 7.3 CDF of downlink sum rate for various NOMA power allocation schemes under OTFS for downlink user average SNR values as indicated below the figures. In fixed power allocation scheme, $\beta_1 = 0.7, \beta_2 = 0.3$. Markers denote results corresponding to various power allocation schemes.

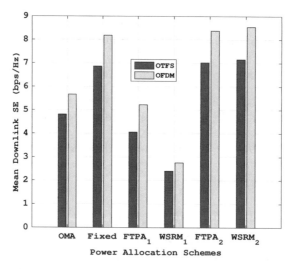

Figure 7.4 Comparative mean SE values for different NOMA power allocation schemes for OTFS and OFDM for downlink user average SNR values $\Gamma_1 = 10$ dB, $\Gamma_2 = 30$ dB. In fixed power allocation scheme, $\beta_1 = 0.7, \beta_2 = 0.3$. For brevity, $\text{FTPA}_1 \equiv$ Avg. SNR-based FTPA, $\text{FTPA}_2 \equiv$ Channel norm-based FTPA, $\text{WSRM}_1 \equiv$ Avg. SNR-based WSRM, $\text{WSRM}_2 \equiv$ Inst. CSI-based WSRM.

SNR based FTPA power allocation with respect to OMA-OTFS when $(\Gamma_1, \Gamma_2) = (15, 25)$ dB. The gain is even higher for power allocation schemes like channel-norm based FTPA and instantaneous CSI-based WSRM NOMA-OTFS schemes. For example, when (Γ_1, Γ_2)

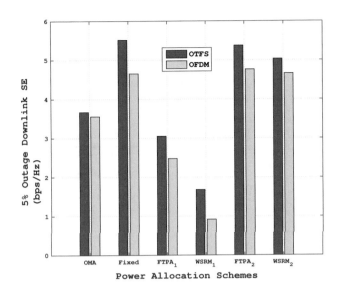

Figure 7.5 Comparative 5% outage SE values for different NOMA power allocation schemes for OTFS and OFDM for downlink user average SNR values $\Gamma_1 = 10$ dB, $\Gamma_2 = 30$ dB. In fixed power allocation scheme, $\beta_1 = 0.7, \beta_2 = 0.3$. For brevity, $FTPA_1 \equiv$ Avg. SNR-based FTPA, $FTPA_2 \equiv$ Channel norm-based FTPA, $WSRM_1 \equiv$ Avg. SNR-based WSRM, $WSRM_2 \equiv$ Inst. CSI-based WSRM.

Table 7.2 Mean and 5% outage SE (in bps/Hz) for downlink NOMA in OTFS and OFDM for $\Gamma_1 = 15$ dB, $\Gamma_2 = 25$ dB for user velocity = 500 kmph.

NOMA powerallocation schemes	Mean SE			5% Outage SE		
	OTFS	OFDM	% gain	OTFS	OFDM	% gain
OMA	4.7618	5.5852	-14.74%	3.931	3.544	10.92%
Fixed-I ($\beta_1 = 0.7, \beta_2 = 0.3$)	5.9499	6.2898	-5.40%	4.925	3.9	26.28%
FTPA (Avg SNR)	5.5487	6.1500	-9.77%	4.574	3.821	19.70%
WSRM (Avg SNR)	3.496	4.0838	-14.39%	2.574	1.658	55.24%
FTPA (channel norm)	5.9977	6.3075	-4.91%	4.874	3.823	27.46%
WSRM (Full CSI)	6.0254	6.2922	-4.24%	4.617	3.654	26.35%

= (10, 30) dB, the mean and 5% outage sum rate shows an increase in the order of $\sim 45\%$ in channel norm based FTPA scheme compared to OMA scheme.

7.5.1.2 Comparison between OTFS and OFDM Performances
Along with the results shown in Table 7.2 and the above mentioned figures, Figure 7.6a and 7.6b, are further included to provide a more objective comparison.

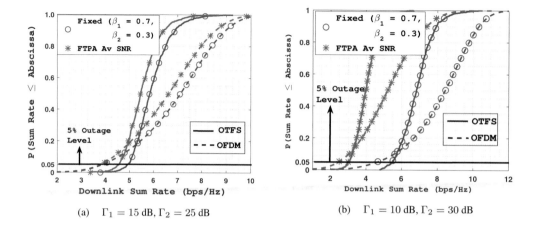

(a) $\Gamma_1 = 15$ dB, $\Gamma_2 = 25$ dB   (b) $\Gamma_1 = 10$ dB, $\Gamma_2 = 30$ dB

Figure 7.6 CDF of downlink sum rate for fixed and average SNR-based FTPA NOMA power allocation schemes under OTFS and OFDM for downlink user average SNR values as in figures, for user velocity = 500 Kmph. Solid and dashed lines represent OTFS and OFDM results respectively. Markers denote results, corresponding to various power allocation schemes.

Table 7.3 Mean and 5% outage SE (in bps/Hz) for uplinklink NOMA in OTFS and OFDM with $\Gamma_1^u = 10$ dB, $\Gamma_2^u = 30$ dB.

Metric	OTFS-NOMA	OFDM-NOMA	% Gain
Mean SE	7.1716	8.6769	-17.34%
5% Outage SE	6.199	5.532	12.05%

It can be observed that the OTFS has significantly higher 5% outage sum SE shows than OFDM. For example, for $(\Gamma_1, \Gamma_2) = (15, 25)$ dB, in case of average SNR based FTPA and channel norm based FTPA schemes, an improvement of around 19.7% and 27.5%, respectively is observed. The gain is higher for weighted sum rate maximization schemes, for which it reaches to ~ 26 % and 55%, respectively. Similar improvement (around 10%) has been observed for OMA scheme as well, highlighting the utility of OTFS over OFDM even in orthogonal multi-user scenario. *The outage improvement in OTFS over OFDM reflects the diversity gain of OTFS with respect to OFDM.*

However, it may be noted that OFDM-based NOMA has nominally higher mean SE (in the order of 5–14%) than OTFS-based NOMA.

Similarly, in uplink NOMA, OTFS is found to have improvement in 5% outage sum rate ($\sim 12\%$) and reduction in mean sum rate ($\sim 17\%$) compared to OFDM when $\Gamma_1^u = 10$ dB and $\Gamma_2^u = 30$ dB and shown in Figure 7.7 and and captured in Table 7.3.

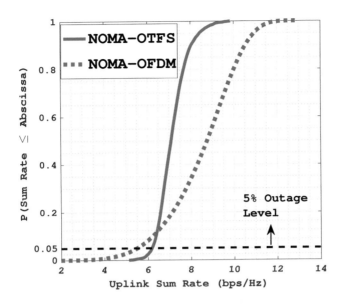

Figure 7.7   CDF of uplink sum rate for NOMA users under OTFS/OFDM for $\Gamma_1^u = 10$ dB, $\Gamma_2^u = 30$ dB for user velocity = 500 Kmph.

7.5.1.3 Comparison of Various NOMA Power Allocation Schemes

From the above depicted results for various NOMA power allocation important observations can be made.

The average SNR-based weighted sum rate maximization scheme's (described in Sec. 7.3.3.1) performance is seen to be the worst compared to the other schemes. This is mainly due to the fact that strong user's average received SNR (Γ_2) is significantly higher than the weak user's average received SNR (Γ_1), thus, resulting in allocation of full power to the weak user (see (7.27)), effectively turning the scheme to OMA with only weak user. Judicious choice of weights (w_i) incorporating proportional fairness can be used to alleviate the issue. The average SNR-based FTPA scheme (described in Sec. 7.3.2.1) also gets partially affected due to the same issue and, thus, the scheme marginally outperforms the OMA scheme. The channel norm based FTPA (described in Sec. 7.3.2.2) and instantaneous CSI-based weighted sum rate maximization (described in Sec. 7.3.3.2) schemes perform significantly better than the OMA scheme in terms of mean and outage spectral efficiency .

7.5.1.4 Extracting NOMA Gain in OTFS with User Channel Heterogeneity

Figure 7.8, shows mean and outage sum rate values for NOMA OTFS with $(\Gamma_1, \Gamma_2)=(15, 25)$ dB and $(10, 30)$ dB. From the figure an interesting observation can be made. We see that both the mean and 5% outage sum rate shows a significant increase for $(\Gamma_1, \Gamma_2)=(10, 30)$ dB compared to $(\Gamma_1, \Gamma_2)=(15, 25)$ dB. Therefore, it can be said that *with an increase in channel heterogeneity (i.e., difference of downlink average SNR) between the user pair who are multiplexed in same De-Do resource block, the downlink sum rate improves due to*

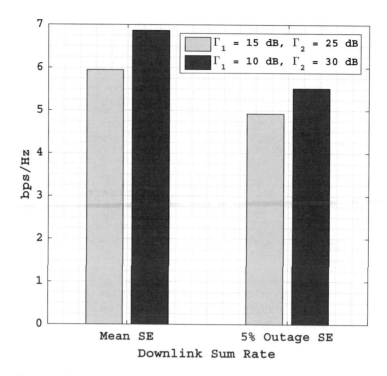

Figure 7.8 Comparative mean and 5% outage sum rate for downlink average SNR pairs $(\Gamma_1, \Gamma_2) = (15, 25)$ dB and (10, 30) dB.

increased NOMA gain. This conforms to well-known observation that increase in channel heterogeneity improves system-level NOMA gain.

7.5.2 Link Level Performance of NOMA-OTFS

System level performance of OTFS-NOMA schemes were described above. In such performance analysis, ideal-SIC implementation was assumed. Whereas in practice such ideal SIC is not realizable. Thus, in this section-link level performance of LDPC coded codeword level SIC NOMA-OTFS system is analysed. Comparison is made against OMA-OTFS and NOMA-OFDM system.

For the purpose of performance evaluation, the average SNR values considered are such that the chosen modulation scheme can experience BLER below the threshold of 10^{-1} [130].

7.5.2.1 Performance of NOMA-OTFS in Downlink

In the downlink direction, each user's data is encoded using LDPC with code rate $R = 2/3$ at base station. The encoded bit stream is modulated using QPSK, 16QAM or 64QAM ($K_i = 2, 4$ and 6, respectively). In order to achieve BLER less than 10^{-1} with LDPC code rate and length shown in Table 7.1. The modulation schemes for both the users are chosen

Table 7.4 SNR threshold values for achieving BLER $\leq 10^{-1}$ for different modulations.

Modulation	SNR(dB)	
	OTFS	OFDM
QPSK	9.5	10.8
16QAM	15.8	18
64QAM	23.5	26

based on the average SINR($\tilde{\Upsilon}$) experienced by the users, which are functions of $\Gamma_1, \Gamma_2, \beta_1$ and β_2 and are given in Table 7.4.

In downlink direction, the average SINR of user 1 is obtained by assuming interference as Gaussian noise, and it is given by,

$$\tilde{\Upsilon}_1 (\text{in dB}) = 10\log_{10}(\frac{\beta_1\Gamma_1}{\beta_2\Gamma_1 + 1}). \qquad (7.44)$$

The post SIC average SINR for user 2, while assuming perfect SIC can be written as:

$$\tilde{\Upsilon}_2 (\text{in dB}) = 10\log_{10}(\beta_2\Gamma_2). \qquad (7.45)$$

The modulated data of the users is transmitted using superposition coding with $\beta_1 = 0.9$ and $\beta_2 = 0.1$ as described in Sec. 7.2.1.1. This choice of β_is results in $\tilde{\Upsilon}_1 = 8.35$ dB for $\Gamma_1 = 15$ dB which puts the system close to minimum operational range. Though the SINR of 8.35 dB is insufficient to satisfy the BLER threshold as per previous discussion but it is observed that the system is able to support required BLER with these β_is. This observation suggests that the Gaussian assumption considered for evaluating SINR may not hold true. User 1 (weak user) decodes the signal using MMSE equalization as outlined in Sec 7.4.1.1. On the other hand, as detailed in Sec. 7.4.1.2, user 2 (strong user) experience higher SNR and thus perform SIC at codeword level. The same SINR thresholds are used for uplink modulation scheme selection as are used for the downlink direction.   Based on the obtained BLER results, throughput (7.42) and goodput (7.43) are computed for each user and plotted in Figure 7.9 and imprinted in Table 7.5.

The results are obtained keeping user 1's modulation as QPSK, as $\Gamma_1 = 15$ dB. Γ_2 is varied such that the higher modulation schemes can be supported by user 2. $\Gamma_2 = 22, 30, 35$ dB are considered. Corresponding to these values of Γ_2, $\tilde{\Upsilon}_2$ obtained from (7.45) support

Table 7.5 Userwise BLER results for downlink NOMA in OTFS and OFDM. ($\beta_1 = 0.9$ and $\beta_2 = 0.1$, code rate= 2/3,UE1 using QPSK Modulation with SNR 15 dB resulting SINR 8.35 dB).

SNR(dB)	SINR(dB)	UE2 Modulation		BLER UE2		BLER UE1		Goodput(bits/s/Hz)		
UE2	UE2	OTFS	OFDM	OTFS	OFDM	OTFS	OFDM	OTFS	OFDM	% gain
22	12	QPSK(2)	QPSK(2)	4.7×10^{-2}	1.2×10^{-1}	3×10^{-3}	1×10^{-3}	2.6	2.51	3.46
25	15	QPSK(2)	QPSK(2)	0	1.3×10^{-3}	1×10^{-3}	3×10^{-3}	2.67	2.66	0.34
30	20	16QAM(4)	16QAM(4)	2×10^{-3}	2.2×10^{-2}	0	6.5×10^{-3}	3.99	3.93	1.50
35	25	64QAM(6)	16QAM(4)	5.6×10^{-2}	3×10^{-4}	5×10^{-3}	5×10^{-3}	5.10	3.99	21.76

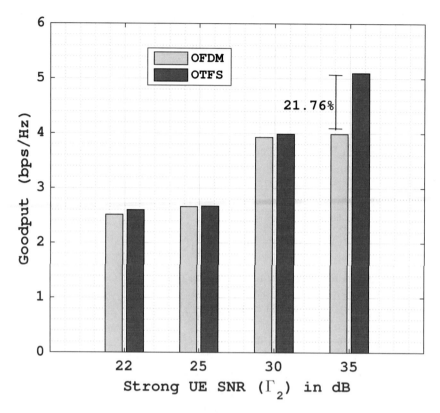

Figure 7.9   Comparative downlink goodput values of NOMA-OTFS and NOMA-OFDM schemes for varied Γ_2 with fixed $\Gamma_1 = 15$ dB.

modulation schemes QPSK,16-QAM and 64-QAM, respectively following the discussion made earlier about the SNR thresholds corresponding to the modulation schemes.

For the SNR pair, $\Gamma_1 = 15$dB and $\Gamma_2 = 35$dB in the table 7.5, user 1 and user 2 are assigned QPSK (K=2) and 64-QAM (K=6)), respectively resulting in a throughput of 5.33 bits/sec/Hz (which is evaluated from (7.42)) for OTFS while the goodput achieved is 5.10 bits/sec/Hz, which is evaluated by taking $K_1 = 2, K_2 = 6, P_{e_1} = 5.6 \times 10^{-2}, P_{e_2} = 5 \times 10^{-3}$ and $R = 2/3$ in (7.43). For the same scenario in OMA case, user 1 can support upto 16QAM while user 2 can support upto 64QAM resulting in throughput of (4*2/3 + 6*2/3)/2 = 3.33 bits/sec/Hz. *Here, the percentage gain in throughput with NOMA-OTFS over OMA-OTFS is 37.52%.*

When, NOMA-OFDM is employed for the same conditions, user 1 is assigned QPSK and user 2 is assigned 16QAM in order to satisfy BLER threshold resulting in a goodput of 3.99 bits/s/Hz. *Thus, NOMA-OTFS offers 21.76% gain in goodput over NOMA-OFDM.*

Table 7.6 Userwise BLER results for uplink NOMA in OTFS and OFDM. (Code rate= 2/3,UE1 using QPSK modulation with SNR 10 dB resulting SINR 10 dB).

SNR(dB)	SINR(dB)	UE2 modulation		BLER UE2		BLER UE1		Goodput(bits/s/Hz)		
UE2	UE2	OTFS	OFDM	OTFS	OFDM	OTFS	OFDM	OTFS	OFDM	% gain
25	15	QPSK(2)	QPSK(2)	0	1.3×10^{-3}	7×10^{-2}	1.8×10^{-1}	2.57	2.42	5.84
30	20	16QAM(4)	16QAM(4)	1×10^{-3}	2.8×10^{-2}	6.7×10^{-2}	3×10^{-1}	3.91	3.52	9.97
40	30	64QAM(6)	64QAM(6)	3.4×10^{-3}	3.6×10^{-2}	9.7×10^{-2}	4.1×10^{-1}	5.19	4.64	10.60

7.5.2.2 Performance of NOMA-OTFS in Uplink

In uplink direction, the table 7.6 is generated by keeping $\Gamma_1^u = 10$ dB and varying the user 2's SNR, $\Gamma_2^u = 25, 30, 40$ dB, thus, varying the user 2's modulation scheme as QPSK,16QAM and 64QAM, respectively. It is easily notable that NOMA using OFDM is unable to support user 1 as P_{e_1} is above threshold. For the SNR pair, $\Gamma_1^u = 10$dB and $\Gamma_2^u = 40$dB, $P_{e_2} = 3.6 \times 10^{-2}$. Thus, it can be seen that user 2 can be supported with *modulation scheme 64QAM but user 1 is unable to transmit even using QPSK due to the* resulting BLER of about 4.1×10^{-1}, which is a result of error propagation. Subsequent values of goodput in the uplink of NOMA-OTFS and NOMA-OFDM systems are illustrated in Figure 7.10. The resulting NOMA-OFDM goodput is 4.64 bits/s/Hz compared with

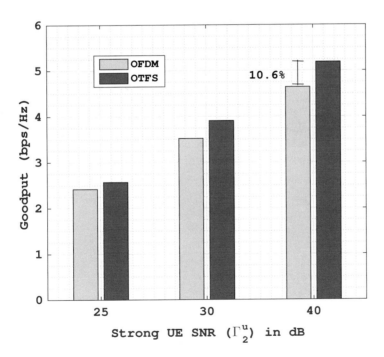

Figure 7.10 Comparative uplink goodput values of NOMA-OTFS and NOMA-OFDM schemes for varied Γ_2^u with fixed $\Gamma_1^u = 10$ dB.

NOMA-OTFS goodput of 5.19 bits/s/Hz, *thus, a gain of 10.60% is shown for NOMA with OTFS over OFDM in uplink.* If OMA is employed for the same scenario, then it can be seen that both user 1 and user 2 can support upto QPSK and 64QAM respectively resulting in a throughput of 2.67 bits/s/Hz, while NOMA throughput is 5.33 bits/s/Hz. *Thus, a gain of 50% in throughput can be obtained in NOMA with respect to OMA.*

7.6 Conclusion

In this chapter, we have presented the a superposition coding based De-Do domain PD-NOMA-OTFS system. Here, NOMA-OTFS is realised using a linear MMSE-SIC receiver.

System level performance analysis using post processing SINR is shown for both downlink and uplink.

A comparison between different partial CSI-based power allocation techniques has been shown.

Link level performance of LDPC coded OTFS-NOMA is also discussed for a 2 user system. Results show that the De-Do domain two-user PD-NOMA-OTFS, as discussed above, has superior performance in comparison with traditional OMA-OTFS by upto 16% in terms of both mean and outage sum SE performance.

It has also been observed that NOMA-OTFS has upto 50% better outage sum SE when compared to NOMA-OFDM for partial CSI-based power allocation schemes. For full CSI-based power allocation schemes, the gain is more than 25%. This also indicates that the OTFS gain over OFDM is not reduced by using NOMA. It is also important to note that mean sum SE of appropriately modified NOMA-OFDM is better than NOMA-OTFS. Thus, we find that there is a tradeoff between mean and outage SE. The improved outage sum SE indicates a robust system.

The link level performance obtained from the developed codeword level SIC receiver shows that the NOMA-OTFS system has upto 21.76% and 10.60% improved goodput in downlink and uplink respectively compared with NOMA-OFDM. It also shows 37.52% and 50% better throughput for NOMA-OTFS over OMA-OTFS system in downlink and uplink, respectively.

A

OTFS Channel Matrix (Ideal)

$$h(\tau, \nu) = \sum_{p=1}^{P} h_p \delta(\tau - \tau_p) \delta(\nu - \nu_p)$$

Now, the received signal can be written as (ignoring the noise term for derivation purpose):

$$r(t) = \int \int h(\tau, \nu) s(t - \tau) e^{j2\pi\nu(t-\tau)} \, d\tau d\nu$$

$$= \sum_{p=1}^{P} h_p s(t - \tau_p) e^{j2\pi\nu_p(t-\tau_p)}$$

Here, $\tau_p = \frac{l_p}{M\Delta f}$ and $\nu_p = \frac{k_p}{NT}$. Therefore, the received signal in discrete domain can be written as:

$$r(n) = \sum_{p=1}^{P} h_p e^{j\frac{2\pi}{MN}k_p(n-l_p)} s\left[(n - l_p)_{MN}\right] \text{ for } n = 0, 1, 2, \cdots, MN - 1.$$

So,

$$r(0) = \sum_{p=1}^{P} h_p e^{j\frac{2\pi}{MN}k_p(-l_p)} s\left[(-l_p)_{MN}\right]$$

$$= h_1 e^{j\frac{2\pi}{MN}k_1(-l_1)} s\left[(-l_1)_{MN}\right] + h_2 e^{j\frac{2\pi}{MN}k_2(-l_2)} s\left[(-l_2)_{MN}\right] + \cdots$$

$$r(1) = \sum_{p=1}^{P} h_p e^{j\frac{2\pi}{MN}k_p(1-l_p)} s\left[(1 - l_p)_{MN}\right]$$

$$= h_1 e^{j\frac{2\pi}{MN}k_1(1-l_p)} s\left[(1 - l_1)_{MN}\right] + h_2 e^{j\frac{2\pi}{MN}k_2(1-l_p)} s\left[(1 - l_2)_{MN}\right] + \cdots$$

$$\vdots$$

Similarly, we can also expand $r(2), r(3), \cdots, r(MN - 1)$.

Now, the received signal vector can be written as:

$$\mathbf{r} = \begin{bmatrix} r(0) \\ r(1) \\ \vdots \\ r(MN-1) \end{bmatrix}$$

$$= h_1 \begin{bmatrix} e^{j\frac{2\pi}{MN}k_1(-l_1)}s\left[(-l_1)_{MN}\right] \\ e^{j\frac{2\pi}{MN}k_1(1-l_1)}s\left[(1-l_1)_{MN}\right] \\ \vdots \\ e^{j\frac{2\pi}{MN}k_1(-1)}s\left[(-1)_{MN}\right] \\ e^{j\frac{2\pi}{MN}k_1(0)}s\left[(0)_{MN}\right] \\ \vdots \\ e^{j\frac{2\pi}{MN}k_1(MN-1-l_1)}s\left[(MN-1-l_1)_{MN}\right] \end{bmatrix}$$

$$+ h_2 \begin{bmatrix} e^{j\frac{2\pi}{MN}k_2(-l_2)}s\left[(-l_2)_{MN}\right] \\ e^{j\frac{2\pi}{MN}k_2(1-l_2)}s\left[(1-l_2)_{MN}\right] \\ \vdots \\ e^{j\frac{2\pi}{MN}k_2(-1)}s\left[(-1)_{MN}\right] \\ e^{j\frac{2\pi}{MN}k_2(0)}s\left[(0)_{MN}\right] \\ \vdots \\ e^{j\frac{2\pi}{MN}k_2(MN-1-l_2)}s\left[(MN-1-l_2)_{MN}\right] \end{bmatrix} + \cdots$$

$$= h_1 \begin{bmatrix} e^{j\frac{2\pi}{MN}k_1(MN-l_1)}s(MN-l_1) \\ e^{j\frac{2\pi}{MN}k_1(MN+1-l_1)}s(MN+1-l_1) \\ \vdots \\ e^{j\frac{2\pi}{MN}k_1(MN-1)}s(MN-1) \\ e^{j\frac{2\pi}{MN}k_1(0)}s(0) \\ \vdots \\ e^{j\frac{2\pi}{MN}k_1(MN-1-l_1)}s(MN-1-l_1) \end{bmatrix} + h_2 \begin{bmatrix} e^{j\frac{2\pi}{MN}k_2(MN-l_2)}s(MN-l_2) \\ e^{j\frac{2\pi}{MN}k_2(MN+1-l_2)}s(MN+1-l_2) \\ \vdots \\ e^{j\frac{2\pi}{MN}k_2(MN-1)}s(MN-1) \\ e^{j\frac{2\pi}{MN}k_2(0)}s(0) \\ \vdots \\ e^{j\frac{2\pi}{MN}k_2(MN-1-l_2)}s(MN-1-l_2) \end{bmatrix} + \cdots$$

$$
= h_1 \Pi^{l_1}
\begin{bmatrix}
e^{j\frac{2\pi}{MN}k_1(0)}s(0) \\
e^{j\frac{2\pi}{MN}k_1(1)}s(1) \\
\vdots \\
e^{j\frac{2\pi}{MN}k_1(MN-1)}s(MN-1)
\end{bmatrix}
+ h_2 \Pi^{l_2}
\begin{bmatrix}
e^{j\frac{2\pi}{MN}k_2(0)}s(0) \\
e^{j\frac{2\pi}{MN}k_2(1)}s(1) \\
\vdots \\
e^{j\frac{2\pi}{MN}k_2(MN-1)}s(MN-1)
\end{bmatrix}
+ \cdots
$$

Here, $\Pi = \begin{bmatrix} 0 & 0 & \cdots & 0 & 1 \\ 1 & 0 & \cdots & 0 & 1 \\ \vdots & \vdots & \ddots & \vdots & \vdots \\ 0 & 0 & \cdots & 1 & 0 \end{bmatrix} = \mathrm{circ}\left\{ \begin{bmatrix} 0 & 1 & 0 & \cdots & 0 \end{bmatrix}_{MN\times1}^T \right\}$ is a circulant delay

matrix of order $MN \times MN$.

$$
\mathbf{r} = h_1 \Pi^{l_1}
\begin{bmatrix}
e^{j\frac{2\pi}{MN}k_1(0)} & 0 & \cdots & 0 \\
0 & e^{j\frac{2\pi}{MN}k_1(1)} & \cdots & 0 \\
\vdots & \vdots & \ddots & \vdots \\
0 & 0 & \cdots & e^{j\frac{2\pi}{MN}k_1(MN-1)}
\end{bmatrix}
\begin{bmatrix}
s(0) \\
s(1) \\
\vdots \\
s(MN-1)
\end{bmatrix}
$$

$$
+ h_2 \Pi^{l_2}
\begin{bmatrix}
e^{j\frac{2\pi}{MN}k_2(0)} & 0 & \cdots & 0 \\
0 & e^{j\frac{2\pi}{MN}k_2(1)} & \cdots & 0 \\
\vdots & \vdots & \ddots & \vdots \\
0 & 0 & \cdots & e^{j\frac{2\pi}{MN}k_2(MN-1)}
\end{bmatrix}
\begin{bmatrix}
s(0) \\
s(1) \\
\vdots \\
s(MN-1)
\end{bmatrix}
+ \cdots
$$

$$
= h_1 \Pi^{l_1}
\begin{bmatrix}
e^{j\frac{2\pi}{MN}(0)} & 0 & \cdots & 0 \\
0 & e^{j\frac{2\pi}{MN}(1)} & \cdots & 0 \\
\vdots & \vdots & \ddots & \vdots \\
0 & 0 & \cdots & e^{j\frac{2\pi}{MN}(MN-1)}
\end{bmatrix}^{k_1}
\begin{bmatrix}
s(0) \\
s(1) \\
\vdots \\
s(MN-1)
\end{bmatrix}
$$

$$
+ h_2 \Pi^{l_2}
\begin{bmatrix}
e^{j\frac{2\pi}{MN}(0)} & 0 & \cdots & 0 \\
0 & e^{j\frac{2\pi}{MN}(1)} & \cdots & 0 \\
\vdots & \vdots & \ddots & \vdots \\
0 & 0 & \cdots & e^{j\frac{2\pi}{MN}(MN-1)}
\end{bmatrix}^{k_2}
\begin{bmatrix}
s(0) \\
s(1) \\
\vdots \\
s(MN-1)
\end{bmatrix}
+ \cdots
$$

$$
= h_1 \Pi^{l_1} \Delta^{k_1} \mathbf{s} + h_2 \Pi^{l_2} \Delta^{k_2} \mathbf{s} + \cdots + h_p \Pi^{l_p} \Delta^{k_p} \mathbf{s}
$$

where, $\Delta = \begin{bmatrix} e^{j\frac{2\pi}{MN}(0)} & 0 & \cdots & 0 \\ 0 & e^{j\frac{2\pi}{MN}(1)} & \cdots & 0 \\ \vdots & \vdots & \ddots & \vdots \\ 0 & 0 & \cdots & e^{j\frac{2\pi}{MN}(MN-1)} \end{bmatrix} = \text{diag}\{\begin{bmatrix} 1 & e^{j\frac{2\pi}{MN}(1)} & \cdots \end{bmatrix}$

$e^{j\frac{2\pi}{MN}(MN-1)}\end{bmatrix}^T\}$ is a diagonal Doppler matrix of order $MN \times MN$ and $\mathbf{s} =$

$\begin{bmatrix} s(0) \\ s(1) \\ \vdots \\ s(MN-1) \end{bmatrix}$ is a transmit signal vector.

Therefore, adding the noise vector $\mathbf{n}$, the received signal vector can be rewritten as:

$$\mathbf{r} = \sum_{p=1}^{P} h_p \Pi^{l_p} \Delta^{k_p} \mathbf{s} + \mathbf{n}$$

$$\implies \mathbf{r} = \mathbf{Hs} + \mathbf{n}$$

Where, $\mathbf{H} = \sum_{p=1}^{P} h_p \Pi^{l_p} \Delta^{k_p}$

References

[1] H. Ochiai, "An analysis of band-limited communication systems from amplifier efficiency and distortion perspective," *IEEE Transactions on Communications*, vol. 61, no. 4, pp. 1460–1472, 2013.

[2] J. Bussgang, L. Ehrman, and J. Graham, "Analysis of nonlinear systems with multiple inputs," *Proceedings of the IEEE*, vol. 62, no. 8, pp. 1088–1119, 1974.

[3] K. Tachikawa, "A perspective on the evolution of mobile communications," *IEEE Communications Magazine*, vol. 41, no. 10, pp. 66–73, Oct. 2003.

[4] K. Murota and K. Hirade, "GMSK Modulation for Digital Mobile Radio Telephony," *IEEE Transactions on Communications*, vol. 29, no. 7, pp. 1044–1050, Jul. 1981.

[5] N. Al-Dhahir and G. Saulnier, "A high-performance reduced-complexity GMSK demodulator," in *Conference Record of The Thirtieth Asilomar Conference on Signals, Systems and Computers*, vol. 1, Nov. 1996, pp. 612–616 vol.1, iSSN: 1058-6393.

[6] H. Liu and K. Li, "A decorrelating RAKE receiver for CDMA communications over frequency-selective fading channels," *IEEE Transactions on Communications*, vol. 47, no. 7, pp. 1036–1045, Jul. 1999.

[7] D. Astely, E. Dahlman, A. Furuskär, Y. Jading, M. Lindström, and S. Parkvall, "LTE: the evolution of mobile broadband," *IEEE Communications Magazine*, vol. 47, no. 4, pp. 44–51, Apr. 2009.

[8] M. Win and Z. Kostic, "Impact of spreading bandwidth on RAKE reception in dense multipath channels," *IEEE Journal on Selected Areas in Communications*, vol. 17, no. 10, pp. 1794–1806, Oct. 1999.

[9] S. S. Das, E. D. Carvalho, and R. Prasad, "Performance Analysis of OFDM Systems with Adaptive Sub Carrier Bandwidth," *IEEE Transactions on Wireless Communications*, vol. 7, no. 4, pp. 1117–1122, Apr. 2008.

[10] G. P. Fettweis, "The tactile internet: Applications and challenges," *IEEE Vehicular Technology Magazine*, vol. 9, no. 1, pp. 64–70, 2014.

[11] 3GPP, *Service requirements for V2X services, TS 22.185, V 14.3, 17-3-2017*, 3GPP, 2017.

[12] GSMA, *The Mobile Economy 2018*, 2018. [Online]. Available: "https://www.gsma.com/mobileeconomy/wp-content/uploads/2018/02/The-Mobile-Economy-Global-2018.pdf"

[13] S. S. Das, E. D. Carvalho, and R. Prasad, "Variable sub-carrier bandwidth in OFDM framework," *Electron. Lett.*, vol. 43, no. 1, Jan. 2007.

[14] S. S. Das, "Techniques to enhance spectral efficiency of OFDM wireless systems," Ph.D. dissertation, Department of Electronic Systems (ES), Aalborg University, Aalborg, 2007.

[15] S. S. Das and I. Rahman, "Method for transmitting data in a wireless network, indian patent, granted, number 260999," Patent 260 999, 2005.

[16] S. S. Das, M. I. Rahman, and F. H. P. Fitzek, "Multi rate orthogonal frequency division multiplexing," in *IEEE International Conference on Communications, 2005. ICC 2005. 2005*, vol. 4, 2005, pp. 2588–2592 Vol. 4.

[17] 3GPP, *NR Physical channels and modulation (Release 15), TS 38.211, V 15.2*, 3GPP, 2018.

[18] FG-NET-2030-Focus Group on Technologies for Network 2030, "Network 2030 : A blueprint of technology, applications and market drivers towards the year 2030 and beyond," *ITU White paper*, pp. 1–19, May 2019. [Online]. Available: https://www.itu.int/en/ITU-T/focusgroups/net2030/Documents/White_Paper.pdf

[19] ——, "Representative use cases and key network requirements for network 2030," January 2020. [Online]. Available: https://www.itu.int/dms_pub/itu-t/opb/fg/T-FG -NET2030-2020-SUB.G1-PDF-E.pdf

[20] ——, "Network 2030 - Additional Representative Use Cases and Key Network Requirements for Network 2030," June 2020. [Online]. Available: https: //www.itu.int/en/ITU-T/focusgroups/net2030/Documents/Additional_use_cases_a nd_key_network_requirements.pdf

[21] Samsung Global News Room, "Samsung's 6g white paper lays out the company's vision for the next generation of communications technology," December 2020. [Online]. Available: https://news.samsung.com/global/samsungs-6g-white-paper-la ys-out-the-companys-vision-for-the-next-generation-of-communications-technolog y

[22] W. Kozek, "Matched weyl-heisenberg expansions of nonstationary environments," 1996.

[23] W. Kozek and A. Molisch, "Nonorthogonal pulseshapes for multicarrier communications in doubly dispersive channels," *IEEE Journal on Selected Areas in Communications*, vol. 16, no. 8, pp. 1579–1589, Oct. 1998.

[24] G. Matz, *A time-frequency calculus for time-varying systems and nonstationary processes with applications*. PhD Thesis, 2000.

[25] ——, "On non-WSSUS wireless fading channels," *IEEE Transactions on Wireless Communications*, vol. 4, no. 5, pp. 2465–2478, Sep. 2005.

[26] T. Strohmer and S. Beaver, "Optimal OFDM design for time-frequency dispersive channels," *IEEE Transactions on Communications*, vol. 51, no. 7, pp. 1111–1122, Jul. 2003.

[27] O. Christensen, *An introduction to frames and Riesz bases*. Springer, 2003, vol. 7.

[28] H. G. Feichtinger and T. Strohmer, Eds., *Gabor Analysis and Algorithms*. Boston, MA: Birkhäuser Boston, 1998. [Online]. Available: http://link.springer.com/10.100 7/978-1-4612-2016-9

[29] D. Gabor, "Theory of communication. Part 1: The analysis of information," *Journal of the Institution of Electrical Engineers - Part III: Radio and Communication Engineering*, vol. 93, no. 26, pp. 429–441, Nov. 1946.

[30] M. Series, "Guidelines for evaluation of radio interface technologies for imt-advanced," *Report ITU*, no. 2135-1, 2009.

[31] J. Wexler and S. Raz, "Discrete Gabor Expansions," *Signal Process.*, vol. 21, no. 3, pp. 207–220, Oct. 1990. [Online]. Available: http://dx.doi.org/10.1016/0165-1684 (90)90087-F

[32] R. W. Chang, "Synthesis of Band-Limited Orthogonal Signals for Multichannel Data Transmission," *Bell System Technical Journal*, vol. 45, no. 10, pp. 1775–1796, Dec. 1966. [Online]. Available: http://onlinelibrary.wiley.com/doi/10.1002/j.1538-7305. 1966.tb02435.x/abstract

[33] G. Matz, D. Schafhuber, K. Grochenig, M. Hartmann, and F. Hlawatsch, "Analysis, Optimization, and Implementation of Low-Interference Wireless Multicarrier Systems," *IEEE Transactions on Wireless Communications*, vol. 6, no. 5, pp. 1921–1931, May 2007.

[34] B. Le Floch, M. Alard, and C. Berrou, "Coded orthogonal frequency division multiplex [TV broadcasting]," *Proceedings of the IEEE*, vol. 83, no. 6, pp. 982–996, Jun. 1995.

[35] R. Haas and J.-C. Belfiore, "A Time-Frequency Well-localized Pulse for Multiple Carrier Transmission," *Wireless Personal Communications*, vol. 5, no. 1, pp. 1–18, Jul. 1997. [Online]. Available: http://link.springer.com/article/10.1023/A%3A1008 859809455

[36] D. Slepian and H. O. Pollak, "Prolate spheroidal wave functions, fourier analysis and uncertainty #x2014; I," *The Bell System Technical Journal*, vol. 40, no. 1, pp. 43–63, Jan. 1961.

[37] A. Vahlin and N. Holte, "Optimal finite duration pulses for OFDM," *IEEE Transactions on Communications*, vol. 44, no. 1, pp. 10–14, Jan. 1996.

[38] P. Jung and G. Wunder, "The WSSUS Pulse Design Problem in Multicarrier Transmission," *IEEE Transactions on Communications*, vol. 55, no. 10, pp. 1918–1928, Oct. 2007.

[39] J. Mazo, "Faster-than-nyquist signaling," *Bell System Technical Journal, The*, vol. 54, no. 8, pp. 1451–1462, Oct. 1975.

[40] F.-M. Han and X.-D. Zhang, "Wireless multicarrier digital transmission via weyl-heisenberg frames over time-frequency dispersive channels," *IEEE Transactions on Communications*, vol. 57, no. 6, pp. 1721–1733, Jun. 2009.

[41] C. W. Korevaar, A. B. J. Kokkeler, P. T. d. Boer, and G. J. M. Smit, "Spectrum Efficient, Localized, Orthogonal Waveforms: Closing the Gap With the Balian-Low Theorem," *IEEE Transactions on Communications*, vol. 64, no. 5, pp. 2155–2165, May 2016.

[42] M. Salehi and J. Proakis, *Digital Communications*. McGraw-Hill Education, 2007. [Online]. Available: https://books.google.co.in/books?id=HroiQAAACAAJ

[43] S. Sesia, I. Toufik, and M. P. J. Baker, Eds., *LTE - the UMTS long term evolution: from theory to practice*, reprinted ed.   Chichester: Wiley, 2009.

[44] 3GPP, "User equipment radio transmission and reception; technical specification part 1: Range 1 standalone (release 15)," 3GPP, Tech. Rep., 2018.

[45] T. Weiss, J. Hillenbrand, A. Krohn, and F. K. Jondral, "Mutual interference in OFDM-based spectrum pooling systems," in *2004 IEEE 59th Vehicular Technology Conference. VTC 2004-Spring (IEEE Cat. No.04CH37514)*, vol. 4, May 2004, pp. 1873–1877 Vol.4.

[46] R. Zayani, Y. Medjahdi, H. Shaiek, and D. Roviras, "WOLA-OFDM: A Potential Candidate for Asynchronous 5g," in *2016 IEEE Globecom Workshops (GC Wkshps)*, Dec. 2016, pp. 1–5.

[47] Y. Medjahdi, R. Zayani, H. Shaïek, and D. Roviras, "WOLA processing: A useful tool for windowed waveforms in 5g with relaxed synchronicity," in *2017 IEEE International Conference on Communications Workshops (ICC Workshops)*, May 2017, pp. 393–398.

[48] J. Abdoli, M. Jia, and J. Ma, "Filtered OFDM: A new waveform for future wireless systems," in *2015 IEEE 16th International Workshop on Signal Processing Advances in Wireless Communications (SPAWC)*, Jun. 2015, pp. 66–70.

[49] L. Zhang, A. Ijaz, P. Xiao, M. M. Molu, and R. Tafazolli, "Filtered ofdm systems, algorithms, and performance analysis for 5g and beyond," *IEEE Transactions on Communications*, vol. 66, no. 3, pp. 1205–1218, 2018.

[50] B. Saltzberg, "Performance of an Efficient Parallel Data Transmission System," *IEEE Transactions on Communication Technology*, vol. 15, no. 6, pp. 805–811, Dec. 1967.

[51] B. Farhang-Boroujeny and C. H. (George) Yuen, "Cosine Modulated and Offset QAM Filter Bank Multicarrier Techniques: A Continuous-Time Prospect," *EURASIP Journal on Advances in Signal Processing*, vol. 2010, pp. 1–17, 2010. [Online]. Available: http://asp.eurasipjournals.com/content/2010/1/165654

[52] B. Hirosaki, "An Orthogonally Multiplexed QAM System Using the Discrete Fourier Transform," *IEEE Transactions on Communications*, vol. 29, no. 7, pp. 982–989, Jul. 1981.

[53] T. H. S. . M. R. Tero Ihalainen, "Channel equalization in filter bank based multicarrier modulation for wireless communications," *EURASIP Journal on Advances in Signal Processing*, vol. 2007, p. 049389 (2006), 2007. [Online]. Available: https://asp-eurasipjournals.springeropen.com/articles/10.1155/2007/49389

[54] T. Ihalainen, T. H. Stitz, and M. Renfors, "Efficient per-carrier channel equalizer for filter bank based multicarrier systems," in *2005 IEEE International Symposium on Circuits and Systems*, 2005, pp. 3175–3178 Vol. 4.

[55] D. Mattera, M. Tanda, and M. Bellanger, "Frequency-spreading implementation of ofdm/oqam systems," in *2012 International Symposium on Wireless Communication Systems (ISWCS)*, 2012, pp. 176–180.

[56] ——, "Performance analysis of the frequency-despreading structure for ofdm/oqam systems," in *2013 IEEE 77th Vehicular Technology Conference (VTC Spring)*, 2013, pp. 1–5.

[57] "Analysis of an fbmc/oqam scheme for asynchronous access in wireless communications," vol. 2015, no. 23.

[58] H. Lin, M. Gharba, and P. Siohan, "Impact of time and carrier frequency offsets on the fbmc/oqam modulation scheme," *Signal Processing*, vol. 102, pp. 151 – 162, 2014. [Online]. Available: http://www.sciencedirect.com/science/article/pii/S01651 68414001121

[59] D. Mattera, M. Tanda, and M. Bellanger, "Performance analysis of some timing offset equalizers for fbmc/oqam systems," *Signal Processing*, vol. 108, pp. 167 – 182, 2015. [Online]. Available: http://www.sciencedirect.com/science/article/pii/S0 165168414004137

[60] M. Bellanger, "Specification and design of a prototype filter for filter bank based multicarrier transmission," in *2001 IEEE International Conference on Acoustics, Speech, and Signal Processing, 2001. Proceedings. (ICASSP '01)*, vol. 4, 2001, pp. 2417–2420 vol.4.

[61] D. S. Waldhauser, L. G. Baltar, and J. A. Nossek, "Mmse subcarrier equalization for filter bank based multicarrier systems," in *2008 IEEE 9th Workshop on Signal Processing Advances in Wireless Communications*, 2008, pp. 525–529.

[62] A. Ikhlef and J. Louveaux, "An enhanced mmse per subchannel equalizer for highly frequency selective channels for fbmc/oqam systems," in *2009 IEEE 10th Workshop on Signal Processing Advances in Wireless Communications*, 2009, pp. 186–190.

[63] P. Lynch, "The dolph–chebyshev window: A simple optimal filter," *AMS Journals*, vol. 125, no. 4, p. 655–660, 1997. [Online]. Available: https://journals.ametsoc.org/v iew/journals/mwre/125/4/1520-0493_1997_125_0655_tdcwas_2.0.co_2.xml

[64] F. Schaich and T. Wild, "Waveform contenders for 5g — ofdm vs. fbmc vs. ufmc," in *2014 6th International Symposium on Communications, Control and Signal Processing (ISCCSP)*, 2014, pp. 457–460.

[65] N. Michailow, S. Krone, M. Lentmaier, and G. Fettweis, "Bit Error Rate Performance of Generalized Frequency Division Multiplexing," in *2012 IEEE Vehicular Technology Conference (VTC Fall)*, Sep. 2012 (Quebec City, Canada), pp. 1–5.

[66] P. C. Chen, B. Su, and Y. Huang, "Matrix Characterization for GFDM: Low Complexity MMSE Receivers and Optimal Filters," *IEEE Transactions on Signal Processing*, vol. PP, no. 99, pp. 1–1, 2017.

[67] M. Matthé and G. Fettweis, "Conjugate-Root Offset-QAM for Orthogonal Multicarrier Transmission," *arXiv:1504.00126 [cs, math]*, Apr. 2015, arXiv: 1504.00126. [Online]. Available: http://arxiv.org/abs/1504.00126

[68] R. Datta, N. Michailow, M. Lentmaier, and G. Fettweis, "GFDM Interference Cancellation for Flexible Cognitive Radio PHY Design," in *2012 IEEE Vehicular Technology Conference (VTC Fall)*, Sep. 2012 (Quebec City, Canada), pp. 1–5.

[69] M. Matthe, D. Zhang, and G. Fettweis, "Iterative Detection using MMSE-PIC Demapping for MIMO-GFDM Systems," in *European Wireless 2016; 22th European Wireless Conference*, May 2016, pp. 1–7.

[70] M. Matthé, D. Zhang, and G. Fettweis, "Sphere-decoding aided SIC for MIMO-GFDM: Coded performance analysis," in *2016 International Symposium on Wireless Communication Systems (ISWCS)*, Sep. 2016, pp. 165–169.

[71] ——, "Low-Complexity Iterative MMSE-PIC Detection for MIMO-GFDM," *IEEE Transactions on Communications*, vol. 66, no. 4, pp. 1467–1480, Apr. 2018.

[72] A. Farhang, N. Marchetti, and L. E. Doyle, "Low-Complexity Modem Design for GFDM," *IEEE Transactions on Signal Processing*, vol. 64, no. 6, pp. 1507–1518, Mar. 2016.

[73] M. Matthé, L. Mendes, I. Gaspar, N. Michailow, D. Zhang, and G. Fettweis, "Precoded GFDM transceiver with low complexity time domain processing," *EURASIP Journal on Wireless Communications and Networking*, vol. 2016, no. 1, p. 1, May 2016. [Online]. Available: http://jwcn.eurasipjournals.springeropen.com/articles/10.1186/s13638-016-0633-1

[74] N. Michailow, M. Matthé, I. S. Gaspar, A. N. Caldevilla, L. L. Mendes, A. Festag, and G. Fettweis, "Generalized frequency division multiplexing for 5th generation cellular networks," *IEEE Transactions on Communications*, vol. 62, no. 9, pp. 3045–3061, 2014.

[75] M. Matthe, I. Gaspar, D. Zhang, and G. Fettweis, "Reduced Complexity Calculation of LMMSE Filter Coefficients for GFDM," in *Vehicular Technology Conference (VTC Fall), 2015 IEEE 82nd*, Sep. 2015, pp. 1–2.

[76] p. wei, X. G. Xia, Y. Xiao, and S. Li, "Fast DGT Based Receivers for GFDM in Broadband Channels," *IEEE Transactions on Communications*, vol. PP, no. 99, pp. 1–1, 2016.

[77] H. Sari, G. Karam, and I. Jeanclaud, "Frequency-domain equalization of mobile radio and terrestrial broadcast channels," in *1994 IEEE GLOBECOM. Communications: The Global Bridge*, Nov. 1994, pp. 1–5 vol.1.

[78] D. Zhang, M. Matthé, L. L. Mendes, and G. Fettweis, "A Study on the Link Level Performance of Advanced Multicarrier Waveforms Under MIMO Wireless Communication Channels," *IEEE Transactions on Wireless Communications*, vol. 16, no. 4, pp. 2350–2365, Apr. 2017.

[79] G. Strang, *Linear algebra and its applications.* Belmont, CA: Thomson, Brooks/Cole, 2006. [Online]. Available: http://www.amazon.com/Linear-Algebra-Its-Applications-Edition/dp/0030105676

[80] T. D. Mazancourt and D. Gerlic, "The inverse of a block-circulant matrix," *IEEE Transactions on Antennas and Propagation*, vol. 31, no. 5, pp. 808–810, Sep. 1983.

[81] M. K. Simon and M.-S. Alouini, *Digital communication over fading channels.* John Wiley & Sons (Hoboken, NJ, USA), 2005, vol. 95.

[82] I. Gaspar, N. Michailow, A. Navarro, E. Ohlmer, S. Krone, and G. Fettweis, "Low Complexity GFDM Receiver Based on Sparse Frequency Domain Processing," in *2013 IEEE Vehicular Technology Conference (VTC Spring)*, Jun. 2013 (Dresden, Germany), pp. 1–6.

[83] A. Viholainen, T. Ihalainen, T. Stitz, M. Renfors, and M. Bellanger, "Prototype filter design for filter bank based multicarrier transmission," in *Signal Processing Conference, 2009 17th European*, Aug. 2009, pp. 1359–1363.

[84] S. Slimane, "Peak-to-average power ratio reduction of OFDM signals using pulse shaping," in *2000 IEEE Global Telecommunications Conference*, Nov. 2000 (San Francisco, USA), pp. 1412–1416.

[85] R. E. Blahut, *Fast algorithms for signal processing*. New York: Cambridge University Press, 2010.

[86] S. T. Chung and A. Goldsmith, "Degrees of freedom in adaptive modulation: a unified view," *IEEE Transactions on Communications*, vol. 49, no. 9, pp. 1561–1571, Sep. 2001.

[87] H. Holma and A. Toskala, Eds., *LTE for UMTS: Evolution to LTE-Advanced*, second edition ed. Chichester, West Sussex, United Kingdom: Wiley, 2011.

[88] S. Slimane, "Peak-to-average power ratio reduction of OFDM signals using broadband pulse shaping," in *2002 IOEEE Vehicular Technology Conference VTC Fall*, vol. 2, 2002, pp. 889–893 vol.2.

[89] H. Myung, J. Lim, and D. Goodman, "Peak-To-Average Power Ratio of Single Carrier FDMA Signals with Pulse Shaping," in *2006 IEEE International Symposium on Personal, Indoor and Mobile Radio Communications(PIMRC)*, Sep. 2006 (Helsinki, Finland), pp. 1–5.

[90] M. Matthe, N. Michailow, I. Gaspar, and G. Fettweis, "Influence of pulse shaping on bit error rate performance and out of band radiation of Generalized Frequency Division Multiplexing," in *2014 IEEE International Conference on Communications Workshops (ICC)*, Jun. 2014 (Sydney, Australia), pp. 43–48.

[91] R. Hadani, S. Rakib, A. F. Molisch, C. Ibars, A. Monk, M. Tsatsanis, J. Delfeld, A. Goldsmith, and R. Calderbank, "Orthogonal Time Frequency Space (OTFS) modulation for millimeter-wave communications systems," in *2017 IEEE MTT-S International Microwave Symposium (IMS)*, Jun. 2017, pp. 681–683.

[92] P. Raviteja, Y. Hong, E. Viterbo, and E. Biglieri, "Practical Pulse-Shaping Waveforms for Reduced-Cyclic-Prefix OTFS," *IEEE Transactions on Vehicular Technology*, vol. 68, no. 1, pp. 957–961, Jan. 2019.

[93] P. Raviteja, E. Viterbo, and Y. Hong, "OTFS Performance on Static Multipath Channels," *IEEE Wireless Commun. Lett.*, pp. 1–1, 2019.

[94] S. Tiwari, S. S. Das, and V. Rangamgari, "Low complexity LMMSE Receiver for OTFS," *IEEE Communications Letters*, vol. 23, no. 12, pp. 2205–2209, Dec. 2019.

[95] T. P. Krauss, M. D. Zoltowski, and G. Leus, "Simple MMSE equalizers for CDMA downlink to restore chip sequence: comparison to zero-forcing and RAKE," in *2000 IEEE International Conference on Acoustics, Speech, and Signal Processing. Proceedings (Cat. No.00CH37100)*, vol. 5, Jun. 2000, pp. 2865–2868 vol.5.

[96] Y. Jiang, M. K. Varanasi, and J. Li, "Performance Analysis of ZF and MMSE Equalizers for MIMO Systems: An In-Depth Study of the High SNR Regime," *IEEE Transactions on Information Theory*, vol. 57, no. 4, pp. 2008–2026, Apr. 2011.

[97] M. Mohaisen and K. Chang, "On the achievable improvement by the linear minimum mean square error detector," in *2009 9th International Symposium on Communications and Information Technology*, Sep. 2009, pp. 770–774.

[98] P. Raviteja, K. T. Phan, Y. Hong, and E. Viterbo, "Embedded Delay-Doppler Channel Estimation for Orthogonal Time Frequency Space Modulation," in *2018 IEEE 88th Vehicular Technology Conference (VTC-Fall)*, Aug. 2018, pp. 1–5.

[99] A. Farhang, A. RezazadehReyhani, L. E. Doyle, and B. Farhang-Boroujeny, "Low Complexity Modem Structure for OFDM-Based Orthogonal Time Frequency Space Modulation," *IEEE Wireless Communications Letters*, vol. 7, no. 3, pp. 344–347, Jun. 2018.

[100] K. R. Murali and A. Chockalingam, "On OTFS Modulation for High-Doppler Fading Channels," in *2018 Information Theory and Applications Workshop (ITA)*, Feb. 2018, pp. 1–10.

[101] P. Raviteja, K. T. Phan, Y. Hong, and E. Viterbo, "Interference Cancellation and Iterative Detection for Orthogonal Time Frequency Space Modulation," *IEEE Transactions on Wireless Communications*, vol. 17, no. 10, pp. 6501–6515, Oct. 2018.

[102] R. Hadani and A. Monk, "OTFS: A New Generation of Modulation Addressing the Challenges of 5G," *arXiv:1802.02623 [cs, math]*, Feb. 2018, arXiv: 1802.02623. [Online]. Available: http://arxiv.org/abs/1802.02623

[103] R. Hadani, S. S. Rakib, A. Ekpenyong, C. Ambrose, and S. Kons, "Receiver-side processing of orthogonal time frequency space modulated signals," US Patent US20 190 081 836A1, Mar., 2019. [Online]. Available: https://patents.google.com/patent/US20190081836A1

[104] L. Li, Y. Liang, P. Fan, and Y. Guan, "Low Complexity Detection Algorithms for OTFS under Rapidly Time-Varying Channel," in *2019 IEEE 89th Vehicular Technology Conference (VTC2019-Spring)*, Apr. 2019, pp. 1–5.

[105] D. W. Walker, T. Aldcroft, A. Cisneros, G. C. Fox, and W. Furmanski, "LU Decomposition of Banded Matrices and the Solution of Linear Systems on Hypercubes," in *Proceedings of the Third Conference on Hypercube Concurrent Computers and Applications - Volume 2*, ser. C^3P. New York, NY, USA: ACM, 1988, pp. 1635–1655, event-place: Pasadena, California, USA. [Online]. Available: http://doi.acm.org/10.1145/63047.63124

[106] G. H. Golub and C. F. Van Loan, *Matrix computations*. JHU Press, 2012, vol. 3.

[107] F. Wiffen, L. Sayer, M. Z. Bocus, A. Doufexi, and A. Nix, "Comparison of OTFS and OFDM in Ray Launched sub-6 GHz and mmWave Line-of-Sight Mobility Channels," in *2018 IEEE 29th Annual International Symposium on Personal, Indoor and Mobile Radio Communications (PIMRC)*, Sep. 2018, pp. 73–79, iSSN: 2166-9589, 2166-9570.

[108] "IEEE standard for information technology– local and metropolitan area networks– specific requirements– part 11: Wireless lan medium access control (MAC) and physical layer (PHY) specifications amendment 5: Enhancements for higher throughput," *IEEE Std 802.11n-2009 (Amendment to IEEE Std 802.11-2007 as*

amended by IEEE Std 802.11k-2008, IEEE Std 802.11r-2008, IEEE Std 802.11y-2008, and IEEE Std 802.11w-2009), pp. 1–565, Oct 2009.

[109] Jianguang Zhao, F. Zarkeshvari, and A. H. Banihashemi, "On implementation of min-sum algorithm and its modifications for decoding low-density parity-check (LDPC) codes," *IEEE Transactions on Communications*, vol. 53, no. 4, pp. 549–554, April 2005.

[110] P. Banelli, S. Buzzi, G. Colavolpe, A. Modenini, F. Rusek, and A. Ugolini, "Modulation Formats and Waveforms for 5G Networks: Who Will Be the Heir of OFDM?: An overview of alternative modulation schemes for improved spectral efficiency," *IEEE Signal Processing Magazine*, vol. 31, no. 6, pp. 80–93, Nov. 2014.

[111] G. D. Surabhi, R. M. Augustine, and A. Chockalingam, "On the Diversity of Uncoded OTFS Modulation in Doubly-Dispersive Channels," *IEEE Transactions on Wireless Communications*, vol. 18, no. 6, pp. 3049–3063, Jun. 2019.

[112] 3GPP, "3GPP TS 38.211 v16.3.0 (2020-09), 3rd generation partnership project; technical specification group radio access network; nr; physical channels and modulation (release 16)," 3GPP, Report, September 2020. [Online]. Available: https://portal.3gpp.org/desktopmodules/Specifications/SpecificationDetails.aspx?specificationId=3213

[113] P. Raviteja, K. T. Phan, and Y. Hong, "Embedded Pilot-Aided Channel Estimation for OTFS in Delay–Doppler Channels," *IEEE Transactions on Vehicular Technology*, vol. 68, no. 5, pp. 4906–4917, May 2019.

[114] S. S. Das, F. H. P. Fitzek, E. D. Carvalho, and R. Prasad, "Variable guard interval orthogonal frequency division multiplexing in presence of carrier frequency offset," in *GLOBECOM '05. IEEE Global Telecommunications Conference, 2005.*, vol. 5, 2005, pp. 5 pp.–2941.

[115] 3GPP, *Study on channel model for frequencies from 0.5 to 100 GHz (Release 16)*, TR 38.901, V 16.1, 3GPP, 2020.

[116] M. J. Fernandez-Getino Garcia, J. M. Paez-Borrallo, and S. Zazo, "Dft-based channel estimation in 2d-pilot-symbol-aided ofdm wireless systems," in *IEEE VTS 53rd Vehicular Technology Conference, Spring 2001. Proceedings (Cat. No.01CH37202)*, vol. 2, 2001, pp. 810–814 vol.2.

[117] J. C. Bartels, R. H.; Beatty and B. A. Barsky, *Hermite and Cubic Spline Interpolation Ch. 3 in An Introduction to Splines for Use in Computer Graphics and Geometric Modelling*, 1998.

[118] V. Rangamgari, S. Tiwari, S. S. Das, and S. C. Mondal, "OTFS: Interleaved OFDM with Block CP," in *National Conference on Communications*, Feb. 2020.

[119] V. Khammammetti and S. K. Mohammed, "OTFS-based multiple-access in high doppler and delay spread wireless channels," *IEEE Wireless Communications Letters*, vol. 8, no. 2, pp. 528–531, April 2019.

[120] R. Hadani and S. S. Rakib, "Multiple access in an orthogonal time frequency space communication system," May 8 2018, uS Patent 9,967,758.

[121] M. T. Le *et al.*, "Fundamental limits of low-density spreading noma with fading," *IEEE Transactions on Wireless Communications*, vol. 17, no. 7, pp. 4648–4659, 2018.

[122] D. Tse and P. Viswanath, *Fundamentals of wireless communication.*   Cambridge university press, 2005.

[123] T. M. Cover and J. A. Thomas, *Elements of information theory.*   John Wiley & Sons, 2012.

[124] R. Hadani, S. Rakib, M. Tsatsanis, A. Monk, A. J. Goldsmith, A. F. Molisch, and R. Calderbank, "Orthogonal Time Frequency Space Modulation," in *IEEE WCNC*, Mar. 2017, pp. 1–6.

[125] Ronny Hadani, Shlomo Rakib, Shachar Kons, Michael Tsatsanis, Anton Monk, Christian Ibars, Jim Delfeld, Yoav Hebron, Andrea J. Goldsmith, Andreas F. Molisch, and Robert Calderbank, "Orthogonal time frequency space modulation," in *https://arxiv.org/abs/1808.00519.*

[126] A. Paulraj, R. Nabar, and D. Gore, *Introduction to Space-Time Wireless Communications*, 1st ed.   New York, NY, USA: Cambridge University Press, 2008.

[127] G. Nain, S. S. Das, and A. Chatterjee, "Low complexity user selection with optimal power allocation in downlink NOMA," *IEEE Wireless Communications Letters*, vol. 7, no. 2, pp. 158–161, 2017.

[128] T. Seyama, T. Dateki, and H. Seki, "Efficient selection of user sets for downlink non-orthogonal multiple access," in *2015 IEEE 26th Annual International Symposium on Personal, Indoor, and Mobile Radio Communications (PIMRC).*   IEEE, 2015, pp. 1062–1066.

[129] A. Monk, R. Hadani, M. Tsatsanis, and S. Rakib, "OTFS-orthogonal time frequency space," *arXiv preprint arXiv:1608.02993*, 2016.

[130] K. Saito, A. Benjebbour, Y. Kishiyama, Y. Okumura, and T. Nakamura, "Performance and design of SIC receiver for downlink NOMA with open-loop SU-MIMO," in *2015 IEEE International Conference on Communication Workshop (ICCW).*   IEEE, 2015, pp. 1161–1165.

[131] C. Yan, A. Harada, A. Benjebbour, Y. Lan, A. Li, and H. Jiang, "Receiver design for downlink non-orthogonal multiple access (NOMA)," in *2015 IEEE 81st vehicular technology conference (VTC Spring).*   IEEE, 2015, pp. 1–6.

[132] L. Yuan, J. Pan, N. Yang, Z. Ding, and J. Yuan, "Successive interference cancellation for LDPC coded nonorthogonal multiple access systems," *IEEE Transactions on Vehicular Technology*, vol. 67, no. 6, pp. 5460–5464, June 2018.

[133] ITU, "Guidelines for Evaluation of Radio Interface Technologies for IMT-Advanced," International Telecommunication Union, Geneva, Recommendation M2135, Dec. 2009.

[134] Won Gi Jeon, Kyung Hi Chang, and Yong Soo Cho, "An equalization technique for orthogonal frequency-division multiplexing systems in time-variant multipath channels," *IEEE Transactions on Communications*, vol. 47, no. 1, pp. 27–32, Jan 1999.

[135] S. S. Das, E. D. Carvalho, and R. Prasad, "Performance analysis of OFDM systems with adaptive sub carrier bandwidth," *IEEE Transactions on Wireless Communications*, vol. 7, no. 4, pp. 1117–1122, April 2008.

[136] A. A. Zaidi *et al.*, "Waveform and numerology to support 5G services and requirements," *IEEE Communications Magazine*, vol. 54, no. 11, pp. 90–98, November 2016.

[137] S. Sesia, I. Toufik, and M. Baker, *LTE-the UMTS long term evolution: from theory to practice*. John Wiley & Sons, 2011.

[138] E. Dahlman, S. Parkvall, and J. Skold, *4G: LTE/LTE-advanced for mobile broadband*. Academic press, 2013.

[139] A. A. M. Saleh, "Frequency independent and frequency dependent non-linear models of twt amplifiers," *IEEE Trans. Commun.*, vol. 29, no. 11, pp. 1715–1720, Nov 1981.

[140] C. Rapp, "Effects of hpa-nonlinearity on a 4-dpsk/ofdm-signal for a digital sound broadcasting system," *Proc. of 2nd European Conf. on Satellite communications.*

[141] T. Helaly, R. Dansereau, and M. El-Tanany, "Ber performance of ofdm signals in presence of nonlinear distortion due to sspa," *Wireless Personal Communications*, vol. 64, no. 4, pp. 749–760, 2012.

[142] D.-S. Han and T. Hwang, "An adaptive pre-distorter for the compensation of hpa nonlinearity," *IEEE Transactions on broadcasting*, vol. 46, no. 2, pp. 152–157, 2000.

[143] Y. Guo and J. R. Cavallaro, "Enhanced power efficiency of mobile ofdm radio using predistortion and post-compensation," in *Proceedings IEEE 56th Vehicular Technology Conference*, vol. 1. IEEE, 2002, pp. 214–218.

[144] G. Surabhi, R. M. Augustine, and A. Chockalingam, "Peak-to-average power ratio of otfs modulation," *IEEE Communications Letters*, vol. 23, no. 6, pp. 999–1002, 2019.

[145] S. Wu and Y. Bar-Ness, "Ofdm systems in the presence of phase noise: consequences and solutions," *IEEE Transactions on Communications*, vol. 52, no. 11, pp. 1988–1996, 2004.

[146] S. Wu, "Phase noise effects on ofdm: analysis and mitigation," 2004.

[147] P. Robertson and S. Kaiser, "Analysis of the effects of phase-noise in orthogonal frequency division multiplex (ofdm) systems," in *Proceedings IEEE International Conference on Communications ICC'95*, vol. 3. IEEE, 1995, pp. 1652–1657.

[148] S. Wu and Y. Bar-Ness, "A phase noise suppression algorithm for ofdm-based wlans," *IEEE Communications Letters*, vol. 6, no. 12, pp. 535–537, 2002.

[149] D. Petrovic, W. Rave, and G. Fettweis, "Effects of phase noise on ofdm systems with and without pll: Characterization and compensation," *IEEE Transactions on communications*, vol. 55, no. 8, pp. 1607–1616, 2007.

[150] G. Fettweis, M. Löhning, D. Petrovic, M. Windisch, P. Zillmann, and W. Rave, "Dirty rf: A new paradigm," *International Journal of Wireless Information Networks*, vol. 14, no. 2, pp. 133–148, 2007.

[151] G. Surabhi, M. K. Ramachandran, and A. Chockalingam, "Otfs modulation with phase noise in mmwave communications," in *2019 IEEE 89th Vehicular Technology Conference (VTC2019-Spring)*. IEEE, 2019, pp. 1–5.

[152] V. Rangamgari, S. Tiwari, S. S. Das, and S. C. Mondal, "OTFS: Interleaved OFDM with block CP," in *2020 National Conference on Communications (NCC)*. IEEE, 2020, pp. 1–6.

Index

About the Authors

Suvra Sekhar Das (Member, IEEE) received the B.Eng. degree in electronics and communication engineering from the Birla Institute of Technology, Ranchi, India, and the Ph.D. degree from Aalborg University, Aalborg, Denmark. He was the Senior Scientist of the Innovation Laboratory, Tata Consultancy Services. He is currently Head of G.S. Sanyal School of Telecommunications, IIT Kharagpur, Kharagpur, India. His current research interests include design of waveform, radio access technology, and radio access networks for QoS traffic.

Ramjee Prasad (fellow IEEE), CTIF Global Capsule, Department of Business Development and Technology, also Aarhus University, Herning, Denmark. Dr. Ramjee Prasad, Fellow IEEE, IET, IETE, and WWRF, is a Professor of Future Technologies for Business Ecosystem Innovation (FT4BI) in the Department of Business Development and Technology, Aarhus University, Herning, Denmark. He is the Founder President of the CTIF Global Capsule (CGC). He is also the Founder Chairman of the Global ICT Standardization Forum for India, established in 2009. He has been honored by the University of Rome "Tor Vergata", Italy as a Distinguished Professor of the Department of Clinical Sciences and Translational Medicine on March 15, 2016. He is an Honorary Professor of the University of Cape Town, South Africa, and the University of KwaZulu-Natal, South Africa. He has received the Ridderkorset of Dannebrogordenen (Knight of the Dannebrog) in 2010 from the Danish Queen for the internationalization of top-class telecommunication research and education. He has received several international awards such as IEEE Communications SocietyWireless Communications Technical Committee Recognition Award in 2003 for making a contribution in the field of "Personal,Wireless and Mobile Systems and Networks", Telenor's Research Award in 2005 for impressive merits, both academic and organizational within the field of wireless and personal communication, 2014 IEEE AESS Outstanding Organizational Leadership Award for: "Organizational Leadership in developing and globalizing the CTIF (Center for TeleInFrastruktur) Research Network", and so on. He has been the Project Coordinator of several EC projects, namely, MAGNET, MAGNET Beyond, eWALL. He has published more than 50 books, 1000 plus journal and conference publications, more than 15 patents, over 140 Ph.D. Graduates and a larger number of Masters (over 250). Several of his students are today worldwide telecommunication leaders themselves.